信息技术人才培养系列教材

Java EE

企业级应用开发与实战

Spring+Spring MVC+MyBatis | 微课版

千锋教育 | 策划　方莹 马剑威 | 主编　佟玉军 常娟 陈刚 | 副主编

人民邮电出版社
北京

图书在版编目（CIP）数据

Java EE企业级应用开发与实战 : Spring+Spring MVC+MyBatis : 微课版 / 方莹, 马剑威主编. -- 北京 : 人民邮电出版社, 2022.9
信息技术人才培养系列教材
ISBN 978-7-115-59142-5

Ⅰ. ①J… Ⅱ. ①方… ②马… Ⅲ. ①JAVA语言－程序设计－教材 Ⅳ. ①TP312.8

中国版本图书馆CIP数据核字(2022)第061276号

内 容 提 要

Java 是 Web 企业级开发常用的语言，具有良好的可移植性，可以轻松地运行于多种平台。本书采用基础知识加案例的方式来讲解 SSM 框架的使用方法，通过对基础知识和案例的讲解，使读者掌握相应的知识点，并在最后一章采用一个完整的实战项目带领读者灵活运用前面介绍的各种技术。除此之外，本书的内容紧跟当下的新技术、新知识，使用 IDEA 企业级编辑器，通过 Maven 实战项目的介绍可使读者更加贴近真实的企业级开发过程。

本书可作为高等院校各专业计算机程序设计课程的教材，也可作为 Web 企业级开发爱好者的参考用书。

◆ 主　　编　方　莹　马剑威
　副 主 编　佟玉军　常　娟　陈　刚
　责任编辑　李　召
　责任印制　王　郁　陈　犇
◆ 人民邮电出版社出版发行　　　北京市丰台区成寿寺路 11 号
　邮编　100164　　电子邮件　315@ptpress.com.cn
　网址　https://www.ptpress.com.cn
　北京天宇星印刷厂印刷
◆ 开本：787×1092　1/16
　印张：16.25　　　　2022 年 9 月第 1 版
　字数：481 千字　　　2024 年 7 月北京第 5 次印刷

定价：59.80 元

读者服务热线：(010)81055256　印装质量热线：(010)81055316
反盗版热线：(010)81055315
广告经营许可证：京东市监广登字 20170147 号

前言

如今，科学技术与信息技术的快速发展和社会生产力的变革对 IT 行业从业者提出了新的需求，从业者不仅要具备专业技术能力，还要具备业务实践能力和健全的职业素质——复合型技术技能人才更受企业青睐。党的二十大报告中提到：“全面提高人才自主培养质量，着力造就拔尖创新人才，聚天下英才而用之。”高校毕业生求职面临的第一道门槛就是技能，因此教科书也应紧随时代，根据信息技术和职业要求的变化及时更新。

Java 程序设计是计算机专业的重要专业课。本书从 SSM 框架的基础知识入手，一步步带领读者深入了解 SSM 框架的核心技术。本书融入了作者在实际开发中的领悟和教学经验，将相对复杂的工作流逐步剖析，为读者进一步学习和应用计算机技术奠定良好的基础。

本书在章节编排上循序渐进，在语法阐述上尽量避免使用生硬的术语和枯燥的公式，从项目开发的实际需求入手，将理论知识与实际应用结合，帮助读者学习和成长，快速积累项目开发经验，从而在职场中拥有较高起点。

本书特点

1．案例式教学，理论结合实战

（1）经典案例涵盖所有主要知识点

- ✧ 根据每章重要知识点，精心挑选案例，促进隐性知识与显性知识的转化，将书中隐性的知识外显，或将显性的知识内化。
- ✧ 案例包含运行效果、实现思路、代码详解。案例设置结构清晰，方便教学和自学。

（2）企业级大型项目，帮助读者掌握前沿技术

- ✧ 引入企业一线项目，进行精细化讲解，厘清代码逻辑，从动手实践的角度，帮助读者逐步掌握前沿技术，为高质量就业赋能。

2．立体化配套资源，支持线上线下混合式教学

- ✧ 文本类：教学大纲、教学 PPT、课后习题及答案、测试题库。
- ✧ 素材类：源码包、实战项目、相关软件安装包。
- ✧ 视频类：微课视频、面授课视频。
- ✧ 平台类：教师服务与交流群、锋云智慧教辅平台。

3．**全方位的读者服务，提高教学和学习效率**

✧ 人邮教育社区（www.ryjiaoyu.com）。教师通过社区搜索图书，可以获取本书的出版信息及相关配套资源。

✧ 锋云智慧教辅平台（www.fengyunedu.cn）。教师可登录锋云智慧教辅平台，获取免费的教学和学习资源。该平台是千锋专为高校打造的智慧学习云平台，传承千锋教育多年来在 IT 职业教育领域积累的丰富资源与经验，可为高校师生提供全方位教辅服务，依托千锋先进教学资源，重构 IT 教学模式。

✧ 教师服务与交流群（QQ 群号：777953263）。该群是人民邮电出版社和图书编者一起建立的，专门为教师提供教学服务，分享教学经验、案例资源，答疑解惑，提高教学质量。

致谢及意见反馈

本书的编写和整理工作由高校教师及北京千锋互联科技有限公司高教产品部共同完成，其中主要的参与人员有方莹、马剑威、佟玉军、常娟、陈刚、徐子惠、毕佳豪等。除此之外，千锋教育的 500 多名学员参与了本书的试读工作，他们站在初学者的角度对本书提出了许多宝贵的修改意见，在此一并表示衷心的感谢。

在本书的编写过程中，我们力求完美，但书中难免有一些不足之处，欢迎各界专家和读者朋友给予宝贵的意见，联系方式：textbook@1000phone.com。

编者

2023 年 5 月

目录

第1章 初识 SSM 框架

第2章 Spring 基础

第3章 Spring 中 Bean 的注入

第4章 Spring 中的 AOP

第 5 章 Spring 与数据库的交互

第 6 章 Spring 事务

第 7 章 MyBatis 基础

第 8 章 MyBatis 核心配置

第 9 章 MyBatis 进阶

第 10 章 Spring MVC

第 11 章 Spring MVC 进阶

第 12 章 SSM 框架整合

第13章 敛书网 SSM 框架整合项目

第1章 初识SSM框架

本章学习目标

- 理解SSM框架的基础知识。
- 理解SSH框架与SSM框架的优缺点。
- 掌握Spring、MyBatis和Spring MVC的基本概念。

利用Servlet+JSP开发的网站，功能实现复杂，代码杂乱不易整理，且后期维护比较困难。为解决这些问题，SSI、SSH和SSM等开发框架应运而生。SSM框架是目前企业级开发的主流框架。因此，本章将对SSM框架进行讲解。

1.1 SSM框架概述

SSM是Spring、Spring MVC、MyBatis的缩写，代表Spring、Spring MVC和MyBatis这3个框架。这3个框架分别用于对实体类、数据库交互和请求层进行封装管理。下面分别对Spring、MyBatis和Spring MVC框架进行讲解，此部分内容需要读者初步理解。

1.1.1 Spring框架

在Spring框架出现之前，Java主要使用企业级JavaBean，即EJB（Enterprise JavaBean）进行企业级开发，但是EJB框架过于依赖容器，并且加载速度慢、使用方式复杂。随着需求的不断增加，EJB已经不再适用于传统的企业级开发。

Spring框架的出现解决了加载速度慢和使用方式复杂等问题。Spring框架采用容器的概念，将需要使用的对象放到容器中，这些对象被称为Spring的Bean。当要使用这些Bean对象时，可以直接从容器中取出，不用重新创建。因此解决了对象的冗余问题，并且减少了开发人员的工作量。Spring容器示意图如图1.1所示。

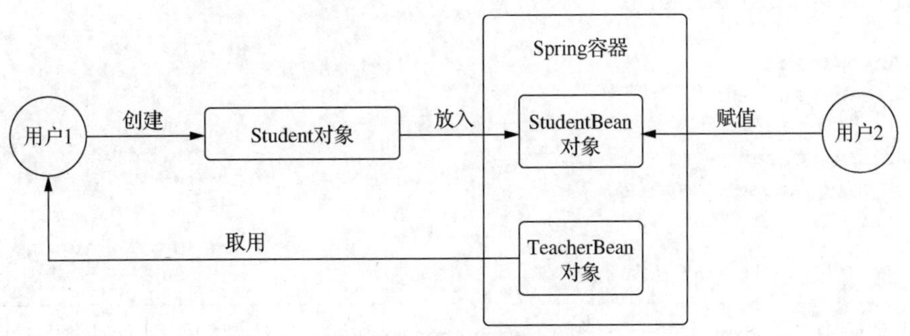

图1.1 Spring容器示意图

从图 1.1 中可以看出，用户可以创建 Student 对象，并将其交由 Spring 容器管理。同样，用户可以取用 Spring 容器中的 Bean 对象或者操纵 Bean 对象的属性。下面通过一个示例为读者展示 Spring 框架的强大之处。

在此使用普通的 Java 业务逻辑处理方式编写一个增加学生分数的示例，代码如下所示。

```
1.  //实体类
2.  public class Student {
3.      Integer id;
4.      Integer score;
5.
6.      public void setScore(Integer score) {
7.          this.score = score;
8.      }
9.  }
10. //业务类
11. public class StudentService {
12.
13.     public Student addScore(){
14.         Student student = new Student();
15.         student.setScore(4);
16.         return student;
17.     }
18. }
```

以上代码中的第 14～16 行创建了一个 Student 对象，并将其分数设置为 4，随后返回该 Student 对象。调用 addScore()方法会新建一个 Student 对象，当此方法被重复调用时，会产生冗余对象。除此之外，Student 与 StudentService 有依赖关系，这不利于后续的维护工作。下面使用 Spring 框架的管理方式修改上述代码，代码如下所示。

```
1.  //实体类
2.  @Component
3.  public class Student {
4.      Integer id;
5.      Integer score;
6.
7.      public void setScore(Integer score) {
8.          this.score = score;
9.      }
10. }
11. //业务类
12. public class StudentService {
13.
14.     @Autowired
15.     Student student;
16.
17.     public Student addScore(){
18.         student.setScore(4);
19.         return student;
20.     }
21. }
```

在以上代码中，Student 类的上方增加了@Component 注解，此注解的作用为创建此类并将其加入 Spring 容器中。业务类代码 student 属性的上方添加了@Autowired 注解，此注解的作用为将容器中对应名称的对象赋给属性。调用业务类中的 addScore()方法后，将容器中 Student 对象的分数设置为 4。

整个过程没有涉及新建的对象，且 Student 与 StudentService 之间没有依赖关系。使用 Spring 框架管理对象降低了类与类之间的耦合性，减少了开发人员的工作量，并且减少了创建对象的数量。

整个 SSM 框架由 Spring 框架管理，Spring 框架掌控并处理容器中所有 Bean 对象的存取，因此 Spring 框架在 SSM 框架中是较为核心的框架。

1.1.2 MyBatis 框架

MyBatis 框架是当前主流的 Java 持久层框架（又称数据传输层框架），因具有高度的灵活性、可优化性和易于维护等特点而被广大互联网厂商青睐。MyBatis 框架为目前企业级开发的首选持久层框架。

MyBatis 框架本质是一个 ORM 框架，ORM 为 Object Relational Mapping 的缩写，即对象关系映射，其主要作用是将数据库中的数据映射到常用的 Java 对象上。ORM 框架的映射流程如图 1.2 所示。

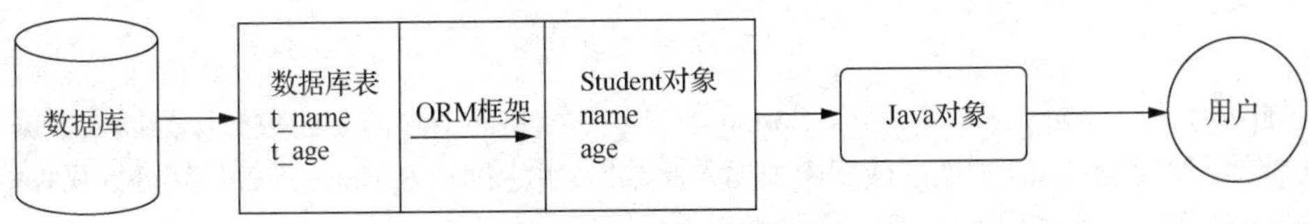

图 1.2 ORM 框架的映射流程

从图 1.2 中可以看出，数据库中的表数据经过 ORM 框架处理后，转化为 Java 对象供用户操作。

MyBatis 框架负责将数据库中查询到的数据封装成 Java 对象，或执行某些数据操作的结构化查询语言（Structured Query Language，SQL）。

1.1.3 Spring MVC 框架

Spring MVC 框架负责处理前端的请求并返回相应的结果，其与 Spring 框架的集成度非常高，可以在 SSM 框架中表现出良好的复用性和扩展性。Spring MVC 框架可以直接使用 Spring 框架的相关功能，并且无需额外的配置文件。

Spring MVC 框架是目前企业级开发常用的 MVC（模型-视图-控制器模式）框架。MVC 是 Model、View 和 Controller 这 3 个模块名称的缩写，它负责规划整个项目的目录结构及业务交互，详细内容将在第 10 章讲解。

在常见的查询操作中，Spring MVC 框架将前端发送过来的请求封装成对象，使用接口的功能处理对象，并根据相应需求通过 MyBatis 框架查询相应的数据，最后将查询结果返回给前端。

1.1.4 SSM 框架的结构

SSM 框架的结构可以分为 3 个部分，分别为 MyBatis 框架、Spring MVC 框架和 Spring 框架。MyBatis 框架负责和数据库进行交互，Spring MVC 框架负责与前端进行交互，Spring 负责掌控整个系统的运行。SSM 框架的结构如图 1.3 所示。

从图 1.3 可以看出，前端发出请求后，此请求由 Spring MVC 框架接收，随后 Spring MVC 框架根据此请求对应的业务决定后续操作。

如果此请求是一个查询操作，那么 Spring MVC 框架将此请求需要查询的相关参数传递给 MyBatis 框架，随后 MyBatis 框架在数据库中查询相关的数据，将返回值封装成 Bean 对象传给 Spring 框架，之后 Spring MVC 框架从 Spring 容器中取出相应的对象返回给前端。

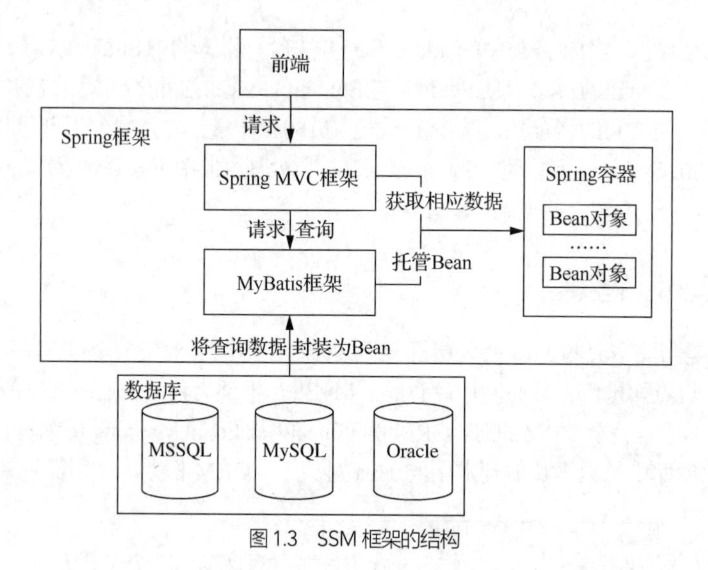

图 1.3 SSM 框架的结构

如果此请求是一个插入操作，那么 Spring MVC 框架将此请求需要的数据作为插入数据传递给 MyBatis 框架，然后 MyBatis 框架向数据库中插入此数据，并返回给 Spring MVC 框架插入操作的状态，随后 Spring MVC 框架将插入状态返回给前端。

1.2 SSH 框架与 SSM 框架的优缺点

Web 应用程序的开发框架有很多，在此通过对比目前流行的 SSH 框架与 SSM 框架，介绍两个框架的优势与不足。

1.2.1 SSH 框架

SSH 框架是 Struts、Spring、Hibernate 这 3 个企业级应用框架名称的首字母缩写。与 SSM 框架中的 Spring、Spring MVC、MyBatis 相比，SSH 框架的不同点在于 Struts 框架和 Hibernate 框架。

SSH 框架将 Hibernate 框架作为数据库持久化层，类似于 SSM 框架中的 MyBatis 框架；将 Struts 框架作为控制层，类似于 SSM 框架中的 Spring MVC 框架。SSH 框架与 SSM 框架均使用 Spring 框架来管理全局对象的流转。

1.2.2 SSH 框架与 SSM 框架对比

1．MyBatis 框架与 Hibernate 框架

MyBatis 框架具有小巧、高效、简单和半自动化等特点。Hibernate 框架的特点是庞大、复杂、间接和全自动化等。

MyBatis 框架对 SQL 的生成较自由，Hibernate 框架对 SQL 的生成追求全自动化。MyBatis 框架具有自由编写 SQL、支持动态 SQL 和支持存储过程等优点，可以灵活定义 SQL 语句，可满足各种开发需求和性能优化的需要。此外，MyBatis 框架的操作较简单，更容易入门。

MyBatis 框架的缺点在于不支持自动化，需要重复编写一些 SQL 语句。但是，随着 MyBatis Plus 等辅助插件的出现，SQL 的生成同样可以实现自动化。因此，在开发方面选用 MyBatis 框架会更便捷。

2．Spring MVC 框架与 Struts 框架

Spring MVC 框架具有简单、便捷、强大等特点，Struts 框架的特点是复杂、烦琐。

在性能方面，Spring MVC 框架实现了低配置，而 Struts 实现了类级别的拦截，每次请求对应实例就要新建一个 Action，加载所有的属性值并注入。所以，Spring MVC 框架的开发效率和性能高于 Struts 框架。

在集成方面，Spring MVC 框架集成了 Ajax，使用方便，只需要一个注解即可完成返回值的设置。此外，Spring MVC 框架与 Spring 框架的契合度较高，可以直接调用 Spring 框架的相关方法，而 Struts 框架需要进行额外的整合配置。因此，通常在开发时选用 Spring MVC 框架。

1.3 本章小结

本章首先讲解了 SSM 框架的基本概念，然后讲解了 SSM 框架和 SSH 框架的优缺点。通过对本章内容的学习，读者可以深入理解 SSM 框架中 3 个框架的概念。

1.4 习题

1．填空题

（1）SSM 是________、________、________的缩写。

（2）在 SSM 框架中，________管理全局对象的流转。

（3）Spring 框架使用________来存放 Bean 对象。

（4）MyBatis 框架在 SSM 框架中负责________与________之间的转换。

（5）Spring MVC 框架用来处理________请求并负责后续的数据流转。

2．选择题

（1）关于 SSM 框架中 MyBatis 框架的功能，下列选项中错误的是（　　）。

A. MyBatis 框架是一种半自动的 ORM 框架

B. 当使用 MyBatis 框架进行开发时，开发人员仍需要编写基本的增删改查语句

C. MyBatis 框架支持动态 SQL

D. MyBatis 框架的入门相较于其他框架更困难

（2）关于 Spring MVC 框架的特点，下列选项中错误的是（　　）。

A. Spring MVC 框架实现了零配置

B. Spring MVC 框架与 Spring 框架的集成度非常高

C. Spring MVC 框架属于 MVC 框架

D. Spring MVC 框架可以直接使用 Spring 框架的相关功能

（3）关于 Spring 框架的相关描述，下列选项中正确的是（　　）。

A. 当程序用到某个对象时，Spring 框架就会将其加入 Spring 容器

B. Spring 框架打破了传统业务中的对象依赖关系，消除了对象之间的耦合性

C. Spring 框架掌控并处理所有 Bean 对象的存取

D. Spring 框架是一个轻量级框架

3．思考题

简述 SSM 框架的优缺点。

第 2 章 Spring 基础

本章学习目标

- 理解 Spring 的基本概念和优势。
- 理解 Spring 的体系结构。
- 理解 Spring 的核心容器。
- 掌握 Bean 对象的创建与获取方法。

Spring 基础

Spring 框架是影响 Java 生态圈最为深远的框架之一，从诞生之初就引起了开发人员的关注。随着 Spring 的不断发展，以 Spring 为核心的企业级开发框架越来越多。Spring 改善了传统重量型框架臃肿、效率低下的缺点，大大降低了项目开发中的技术复杂程度。本章将详细介绍 Spring 的基本概念、优势、体系结构和核心部分，并通过简单示例帮助读者掌握 Bean 对象的创建与获取方法。

2.1 Spring 概述

本节简述 Spring 的基本概念及优势，并对 Spring 的体系结构进行详细的分类总结。通过对本节的学习，读者可初步理解 Spring 的基本概念与优势，了解 Spring 的体系结构。

2.1.1 Spring 简介

Spring 是用于 Java SE 和 Java EE 应用的一站式轻量级开源框架，它最核心的理念是控制反转（Inversion of Control，IoC）和面向切面编程（Aspect Oriented Programming，AOP）。其中，IoC 概念支撑 Spring 对 Bean 对象的管理，而 AOP 技术对 Spring 中事务管理等功能起到关键作用，这些概念分别在 2.2.3 小节和第 4 章做详细介绍。

Spring 提供了展示层、持久层及事务管理等一站式的企业级应用技术。除此之外，Spring 已经融入 Java EE 开发的各个领域，它可以整合 Java 生态圈中众多的第三方框架和类库，为企业开发提供全面的功能支持。

2.1.2 Spring 的优势

与其他开发框架相比，Spring 具有的优势主要体现在以下几个方面。

1．降低耦合度，方便开发

通过 Spring 的核心容器，Spring 框架可以管理对象的生命周期，控制对象之间的依赖关系，以降低对象与对象间的耦合度，使系统易于维护。

2．支持 AOP 编程

Spring 可以通过 AOP 对程序进行权限拦截、安全监控和事务管理等操作，使各部分业务逻辑之间的耦合度降低，提高程序的可重用性。

3．支持声明式事务

事务分为两种，一种是编程式事务，另一种是声明式事务。编程式事务需要手动控制事务的开始与结束，而声明式事务建立在 AOP 之上，只需制订规则即可自动进行事务管理，无须手动控制。Spring 支持声明式事务，可以直接通过 Spring 配置文件来管理数据库事务，提升开发效率。

4．方便程序测试

Spring 集成了 JUnit，开发人员可以通过 JUnit 进行单元测试。

5．方便集成各种优秀框架

Spring 为开发人员提供了广阔的平台，它不排斥各种优秀的开源框架，其内部提供了对各种优秀框架（如 MyBatis、Hibernate、Quartz 等）的直接支持。

6．降低 Java EE API 的使用难度

Spring 封装了 Java EE 中使用难度较大的 API，封装后，这些 API 更容易被开发人员理解和使用。

2.1.3 Spring 的体系结构

Spring 框架采用分层架构，由数据访问、Web 交互层、AOP 框架、核心容器、测试模块和消息模块组成，如图 2.1 所示。

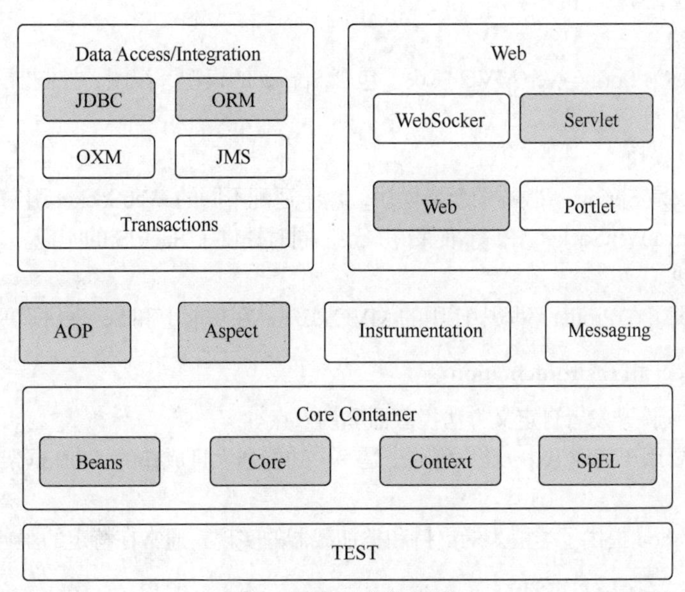

图 2.1 Spring 的体系结构

图中背景为深色的模块需要读者重点了解，下面对这些核心模块进行分类介绍。

1．Data Access/Integration（数据访问及集成）

数据访问及集成主要是对数据流转的控制和操作，主要包含 JDBC、ORM、OXM、JMS 和 Transactions 模块。

（1）JDBC 模块

JDBC 即 Java Database Connectivity（Java 数据库连接），JDBC 模块利用 JDBC 抽象层来间接控制所有的数据库连接代码，大大降低了编程的复杂性。

（2）ORM 模块

ORM 模块即对象关系映射模块，主流的对象关系映射框架包括 MyBatis、Hibernate 和 JDO 等。ORM 模块可将对象关系映射框架与 Spring 提供的 JDBC 模块结合使用。

（3）OXM 模块

OXM 模块与 ORM 模块类似，ORM 模块提供的是数据库数据和对象之间的映射，而 OXM 是 Object XML Mapper 的缩写，译为 XML 对象关系映射。

（4）JMS 模块

JMS 是 Java Message Service 的缩写，译为 Java 消息传递服务。在 Spring 4.1 之后的版本，支持与 Spring-message 模块的集成。

（5）Transaction 模块

Transaction 模块的功能是管理事务，它主要负责管理所有的 POJO 类，同时处理声明式事务和编程式事务。

2．Web

Web 功能的实现基于 ApplicationContext，它提供了 Web 应用的各种工具类，由 Web、Servlet、WebSocket 和 Protlet 模块组成。

（1）Web 模块

Web 模块提供了基本的面向 Web 的集成功能，如文件上传、使用 Servlet 监听器的 IoC 容器的初始化、面向 Web 的应用程序上下文等。

（2）Servlet 模块

Servlet 模块也称为 Spring-web MVC 模块，包含 Spring 的模型、视图、控制器及 Web 应用程序的 REST Web 服务实现。

（3）WebSocket 模块

WebSocket 模块是 Spring 4 版本的新增模块，用于适配不同的 WebSocket 引擎，并能够全面支持 WebSocket。它与 Java WebSocket API 标准保持一致，同时提供了 SockJS 的实现。

（4）Protlet 模块

Protlet 模块提供了在 Protlet 环境中使用的 MVC 实现，在功能上和 Servlet 模块类似。

3．AOP、Aspect 和 Instrumentation

（1）AOP 模块允许开发者自定义方法拦截器和切入点。

（2）Aspect 主要用于集成 AspectJ。Aspect 是一个功能强大且成熟的 AOP 框架，为面向切面编程提供多种实现方法。

（3）Instrumentation 提供了类植入的支持和类加载器的实现，通常在特定的服务器中使用。

4．Core Container（核心容器）

（1）Beans 和 Core 模块

Beans 和 Core 模块是 Spring 中较为核心的模块，提供了 IoC 和 DI 思想的实现，规定了创建、配置和管理 Bean 对象的方式。

（2）Context 模块

Context 模块在 Beans 和 Core 模块的基础之上提供了框架式 Bean 对象的访问方式，通过

ApplicationContext 接口可以获取上下文信息。

（3）SpEL 模块

SpEL 模块提供了强大的表达式语言，该语言用于在 Spring 运行时查询和操作对象。

5．TEST

TEST 模块支持使用 JUnit 或 TestNG 对 Spring 组件进行单元测试和集成测试。它提供了 ApplicationContext 对象和该对象的缓存。除此之外，它还提供了可用于独立测试代码的模拟对象。

6．MESSAGING

Spring 4 版本开始引入 MESSAGING，MESSAGING 主要用于 Spring 框架的消息传送。

2.2 Spring 的核心部分

本节将详细讲述 Spring 的核心部分，主要内容为 IoC 和 DI、Spring 容器和 Spring 中 Bean 对象的概念。本节也是本章的核心部分，读者需要深入理解本节全部内容。

2.2.1 IoC 与 DI

IoC 是一种设计思想，指将对象的控制权交由容器。Spring 采用这种思想后，开发人员无须编写管理依赖关系的代码，由 Spring 容器根据配置自动生成。

DI 为 Dependency Injection（依赖注入）的缩写，是 IoC 思想的一种体现，即通过注入的方式来避免调用类对其他类的强依赖关系，方便开发，减少了代码冗余。下面通过两个示例演示普通依赖和依赖注入，分别如例 2-1 和例 2-2 所示。

【例 2-1】 Student.java

```
1.  public class Student {
2.      Computer computer;
3.
4.      public void setComputer(Computer computer) {
5.          this.computer = computer;
6.      }
7.  }
```

例 2-1 展示的是一个普通的学生类，每个学生拥有一台计算机。在实例化此学生类时，需要创建 Student 类和 Computer 类，并将 Computer 类赋给 Student 类。此时类与类之间具有依赖关系。

【例 2-2】 Student.java

```
1.  @Component
2.  public class Student {
3.      @Autowired
4.      Computer computer;
5.  }
```

例 2-2 展示的是一个被 Spring 管理的学生 Bean 对象（添加@Component 注解即可将此类放入 Spring 容器），当需要实例化此类时，从容器中取用即可，无需其他操作。在此过程中，Spring 容器会扫描到 computer 属性上面的@Autowired 注解，检索容器中是否有 Computer 类，此时为 Computer 类加上@Component 注解，Spring 将从容器中找到 Computer 类并将其自动注入 Student 类的 computer 属性，无需手动赋值。详细注入过程将在第 3 章讲述。

2.2.2 Spring 容器

Spring 容器是根据 IoC 思想构建的，因此 Spring 容器又称为 Spring IoC 容器。Spring 容器有多种实现接口，其中，重要的实现接口有 BeanFactory 和 ApplicationContext。

1．BeanFactory

BeanFactory 是一个接口，它代表 Spring 的抽象容器，其中封装了许多与 Bean 有关的抽象方法，具体代码如例 2-3 所示。

【例 2-3】 BeanFactory.java

```
1.  public interface BeanFactory {
2.      //获取需要的 Bean
3.      Object getBean(String var1) throws BeansException;
4.      //判断容器中是否有此 Bean
5.      boolean containsBean(String var1);
6.      //判断此 Bean 是否是单例
7.      boolean isSingleton(String var1);
8.      //判断此 Bean 是否是多例
9.      boolean isPrototype(String var1);
10.     //获取此 Bean 的类型
11.     @Nullable
12.     Class<?> getType(String var1);
13. }
```

从例 2-3 的代码中可以看出，BeanFactory 中存在获取 Bean 信息的方法和判断 Bean 是否为单例的方法。

2．ApplicationContext

ApplicationContext 是 BeanFactory 的子接口，又称作应用上下文，是重要的 Spring 应用容器。它包含了 BeanFactory 中所有的功能，并且支持国际化、资源访问、事件传播等，是开发中经常使用的接口。

ApplicationContext 的子类 ClassPathXmlApplicationContext 是开发中常用的工具类。它可以通过指定一个配置文件来完成容器的创建，配置文件中通常有配置完成的 Bean 对象。容器创建完成后，其中的 Bean 对象也创建完毕，开发人员可以通过 Bean 对象的名称获取容器中的 Bean 对象、Bean 的类型等信息。创建容器的语法如下所示。

```
ClassPathXmlApplicationContext classPathXmlApplicationContext =
   new ClassPathXmlApplicationContext("配置文件的类路径");
```

上述代码中，ClassPathXmlApplicationContext 是 Spring 容器实现类中的一个。其他 ApplicationContext 的实现类不常用，感兴趣的读者可以自行参考相关读物进行学习。

2.2.3 Spring 中的 Bean

Spring 中的 Bean 对象就是被 Spring 管理的对象。例如，创建 Student 学生对象，此对象一旦被 Spring 管理，就将其称为 Bean 对象。通常 Bean 对象在容器加载时被加入容器，容器销毁时才会被销毁。

在容器的整个运行过程中，一般不会再向容器中添加 Bean 对象。初次学习的读者可以认为容器在加载完成后，其中的 Bean 对象就被固定，开发人员仅可在容器运行期间对 Bean 对象进行取用、赋值和查看操作。

2.3 Spring 示例

本节将初步介绍 Spring 工程搭建的方法，整个过程可分为 Spring 依赖的下载、Web 环境搭建、Bean 对象的创建与获取。读者需要掌握本节全部内容。

2.3.1 Spring 依赖的下载

本书示例采用 Spring 5.0.8，读者需要在官方网站下载对应的 JAR 包。下面讲解 SpringJAR 包的获取方法。

（1）访问 Spring JAR 包的官方下载页面，如图 2.2 所示。

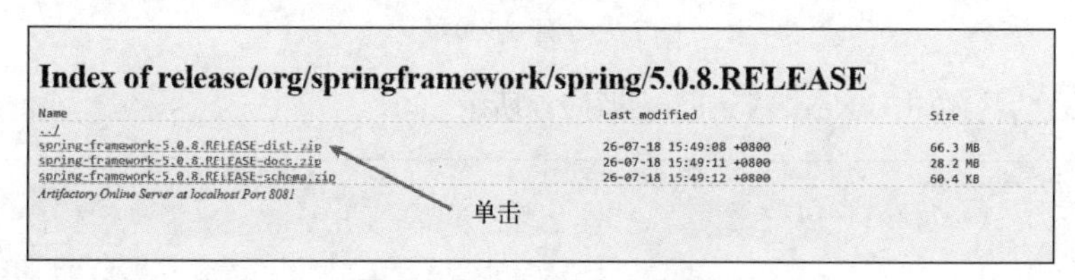

图 2.2 Spring JAR 包下载页面

（2）单击 spring-framework-5.0.8.RELEASE-dist.zip，开始下载依赖包。

（3）访问 Commons Logging 包的官方下载页面，如图 2.3 所示。

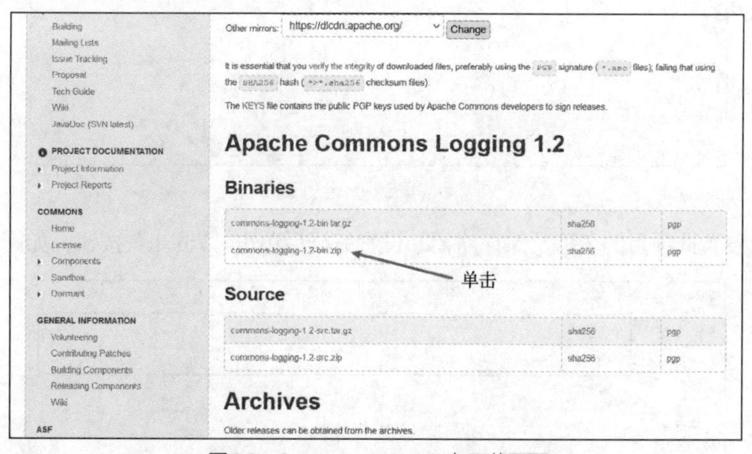

图 2.3 Commons Logging 包下载页面

（4）单击 commons-logging-1.2-bin.zip，开始下载依赖包。

（5）在官方网站下载 AOP 注解依赖的 JAR 包。

2.3.2 Web 环境搭建

Web 环境搭建属于 Java Web 的基础知识，但考虑到部分读者是初次使用 IDEA 编辑器，在此讲述

Web 环境搭建步骤。本示例使用 IDEA 2021.2 进行讲解。

1．创建项目

打开 IDEA，单击“New Project”按钮创建项目，在弹出的对话框中选择 Java 项目，随后单击“Next”按钮，如图 2.4 所示。

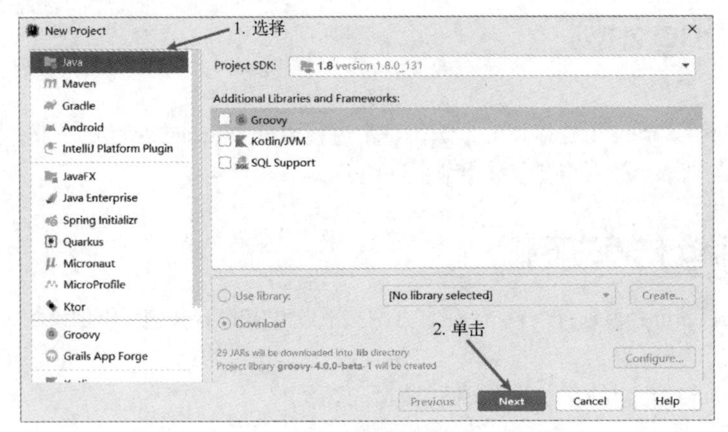

图 2.4　创建 Java 项目

进入下一步之后，单击“Next”按钮，如图 2.5 所示。

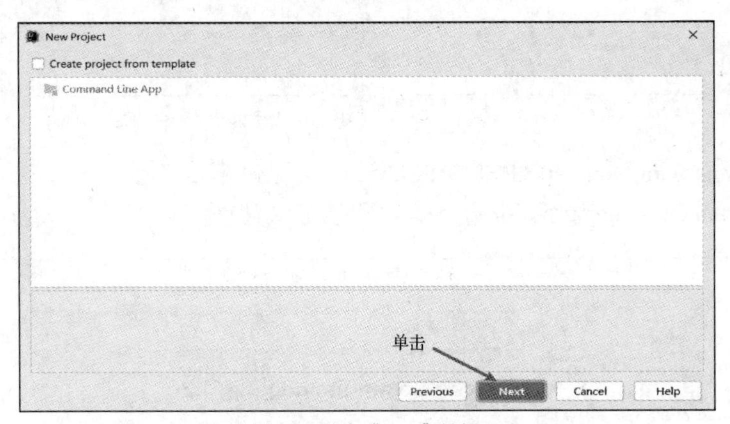

图 2.5　单击“Next”按钮

最后一步，设置此项目的名称及保存路径，设置完毕后单击“Finish”按钮，如图 2.6 所示。

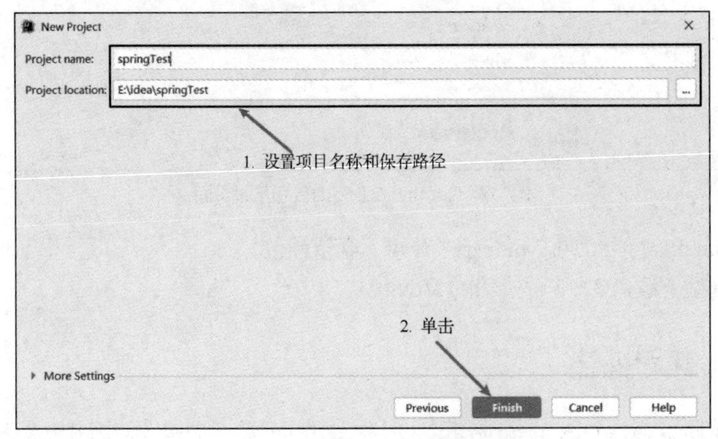

图 2.6　设置项目名称及保存路径

2．为项目添加 Web 框架支持

项目创建完成后，右键单击该项目，在弹出的快捷菜单中选择 Add Framework Support 命令，如图 2.7 所示。

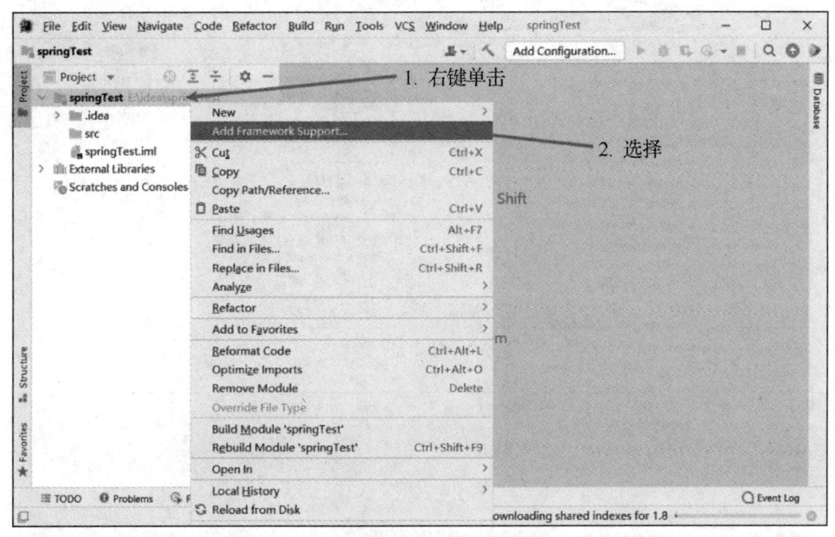

图 2.7　为项目添加框架支持

在弹出的对话框中勾选 Web Application（4.0）及 Create web.xml 复选框，单击“OK”按钮完成添加，如图 2.8 所示。

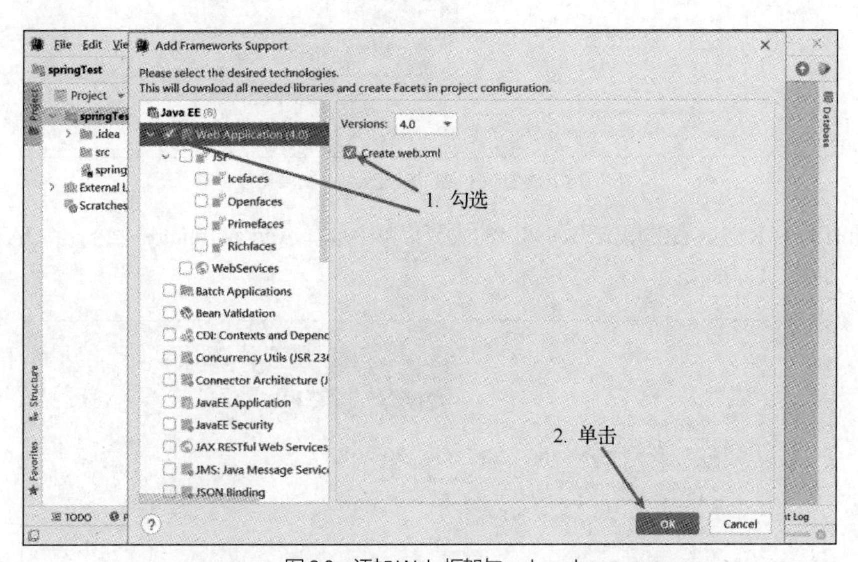

图 2.8　添加 Web 框架与 web.xml

3．添加依赖包

右键单击 WEB-INF 文件夹，在弹出的快捷菜单中选择 New→Directory 命令，创建 lib 文件夹，如图 2.9 所示。

将 2.3.1 小节下载的 Spring 包中 libs 文件夹下的所有 JAR 包、Commons Logging 包和 AOP 注解依赖包复制到 lib 文件夹下，如图 2.10 所示。

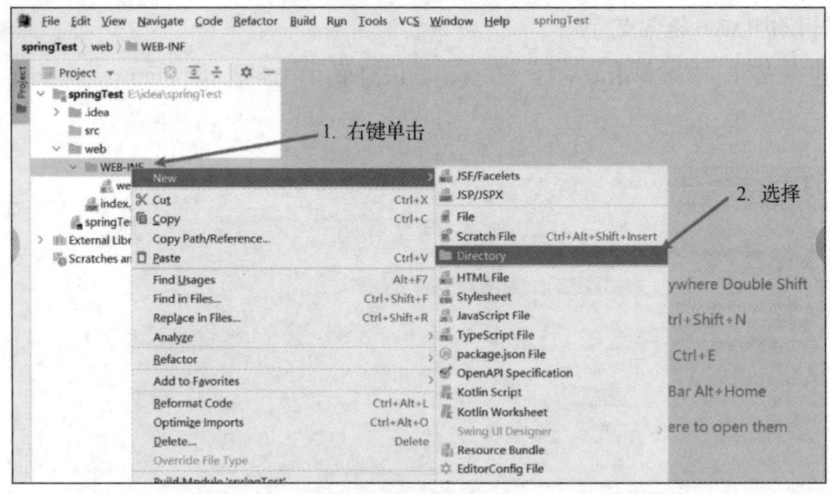

图 2.9　创建 lib 文件夹

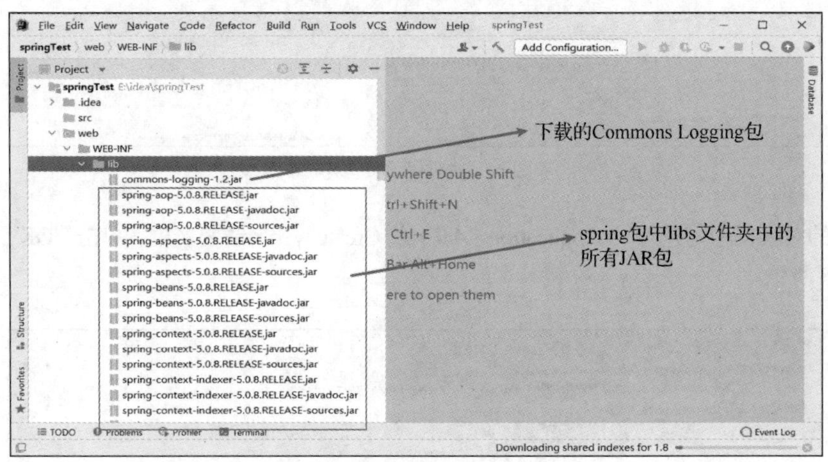

图 2.10　复制需要的依赖包到 lib 文件夹

选择所有的 JAR 包，右键单击，在弹出的快捷菜单中选择 Add as Library 命令，将 JAR 包添加到 Library 中，如图 2.11 所示。

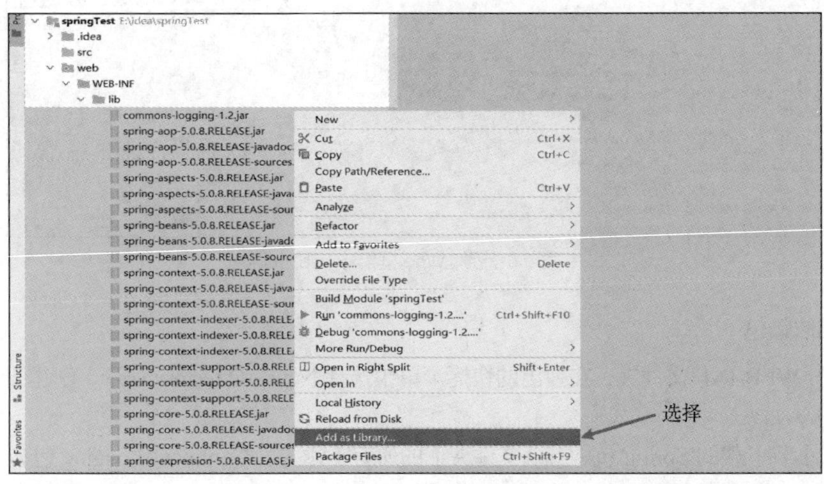

图 2.11　添加 JAR 包到 Library 中

选择 File→Project Structure 命令，打开项目结构对话框，如图 2.12 所示。

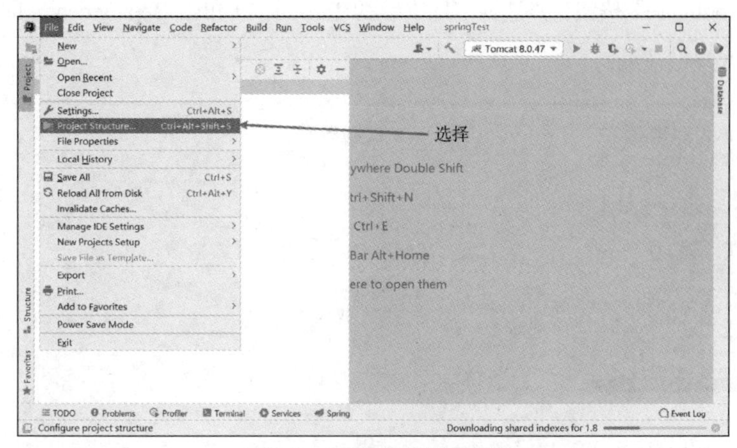

图 2.12　打开项目结构对话框

在项目结构对话框中单击 Artifacts，选择发布包，随后在对话框右侧双击 Commons Logging 包，将其添加到现有项目中，最后单击“Apply”按钮将此包应用到项目中，如图 2.13 所示。

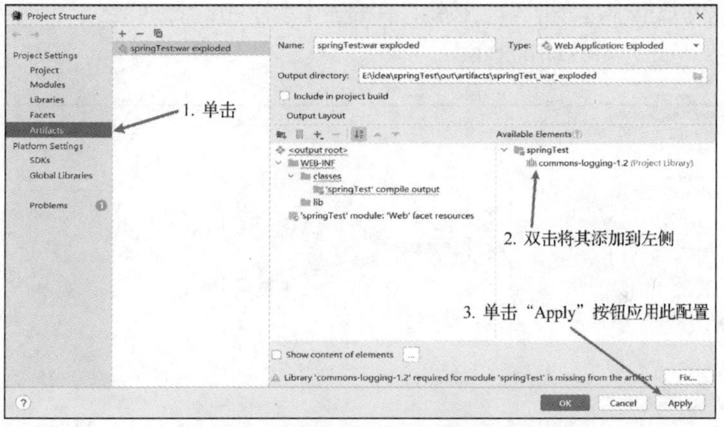

图 2.13　将依赖包应用到项目中

4. 配置 tomcat

单击操作栏中的下拉按钮，在下拉列表中选择 Edit Configurations 选项，打开配置对话框，如图 2.14 所示。

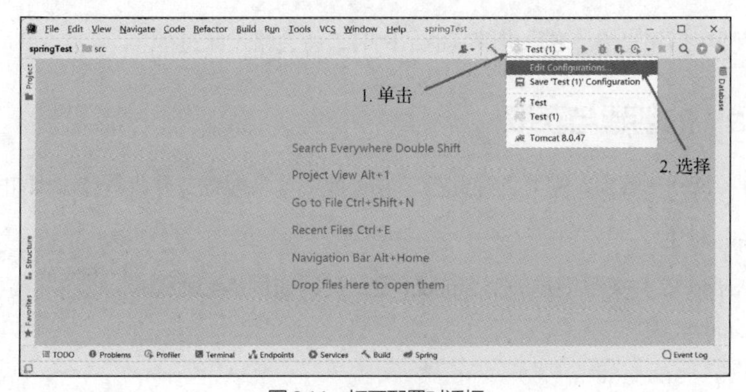

图 2.14　打开配置对话框

单击对话框左上角的“+”按钮，在弹出的菜单中选择 Tomcat Server→Local 命令，在右侧配置 Tomcat。单击“Configure”按钮，并选择本地安装的 Tomcat，再单击 Deployment，即可打开 war 包配置对话框，如图 2.15 所示。

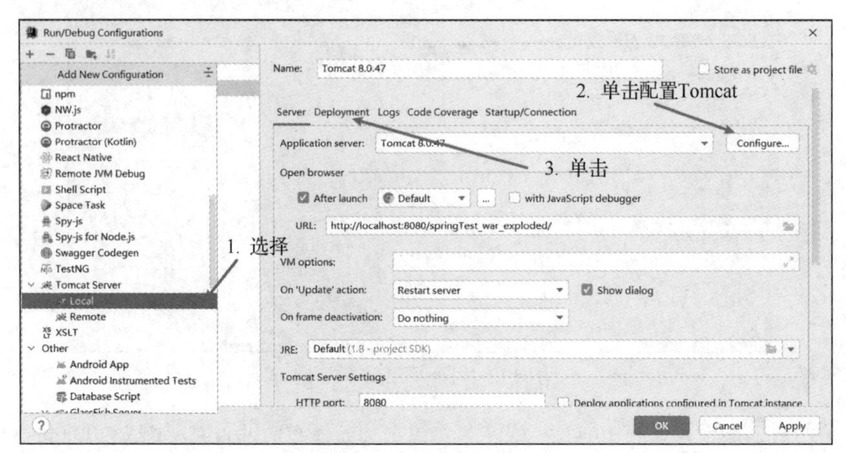

图 2.15 配置 Tomcat

在 war 包配置对话框中，单击“+”按钮，在列表中选择 Artifact 选项，将 war 包添加进 Tomcat，随后单击“Apply”按钮，如图 2.16 所示。

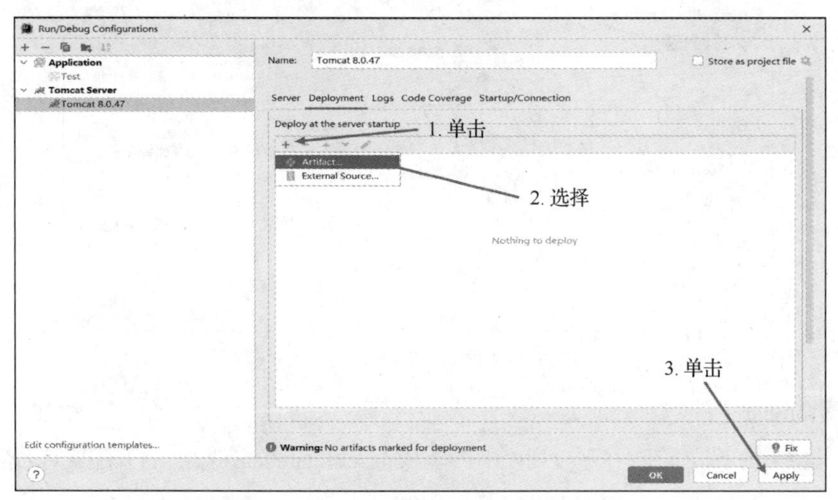

图 2.16 发布 war 包

至此，Web 环境搭建完毕。之后讲解其他示例时，将会略过此过程。

2.3.3 Bean 的创建与获取

将一个类装入 Spring 容器，需要创建此类，并将此类写入配置文件，具体步骤如下。

1. 创建 Bean 对象

在工程文件的 src 文件夹下创建一个 Student 类，代码如例 2-4 所示。

【例 2-4】 Student 类

```
1.  public class Student {
```

```
2.      String name = "张三";
3.      int age = 18;
4.
5.      public Student() {
6.      }
7.      public Student(String name, int age) {
8.          this.name = name;
9.          this.age = age;
10.     }
11.     public String getName() {
12.         return name;
13.     }
14.     public void setName(String name) {
15.         this.name = name;
16.     }
17.     public int getAge() {
18.         return age;
19.     }
20.     public void setAge(int age) {
21.         this.age = age;
22.     }
23.     public String toString() {
24.         return "Student{" +
25.                 "name='" + name + '\'' +
26.                 ", age=" + age +
27.                 '}';
28.     }
29. }
```

2．编写配置文件

右键单击 src 目录，在弹出的快捷菜单中选择 New→XML Configuration File→Spring Config 命令，如图 2.17 所示，创建 Spring 配置文件，将其命名为 applicationContext。

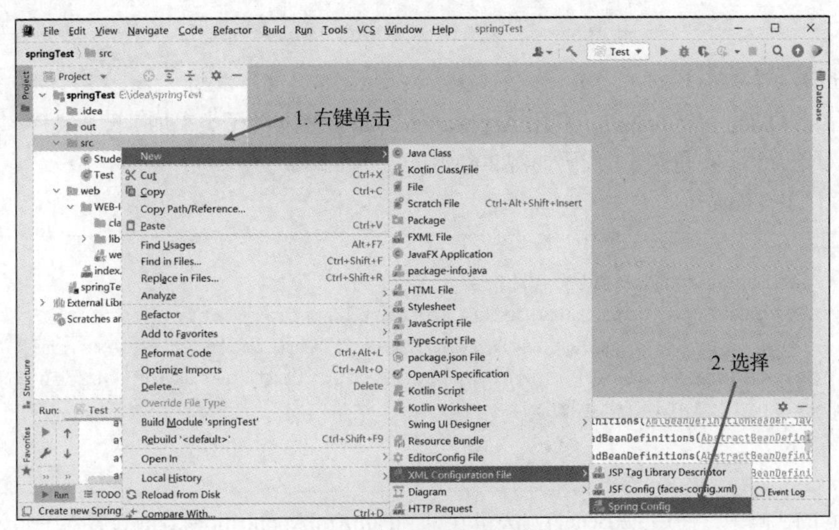

图 2.17　创建 Spring 配置文件

创建完毕后，打开目录结构，单击 Modules，选择 springTest→Spring 选项，之后单击对话框右侧的“+”按钮，在弹出的对话框中勾选新建的 applicationContext.xml 文件，单击“OK”按钮，再单击“Apply”按钮，完成配置操作，如图 2.18 所示。

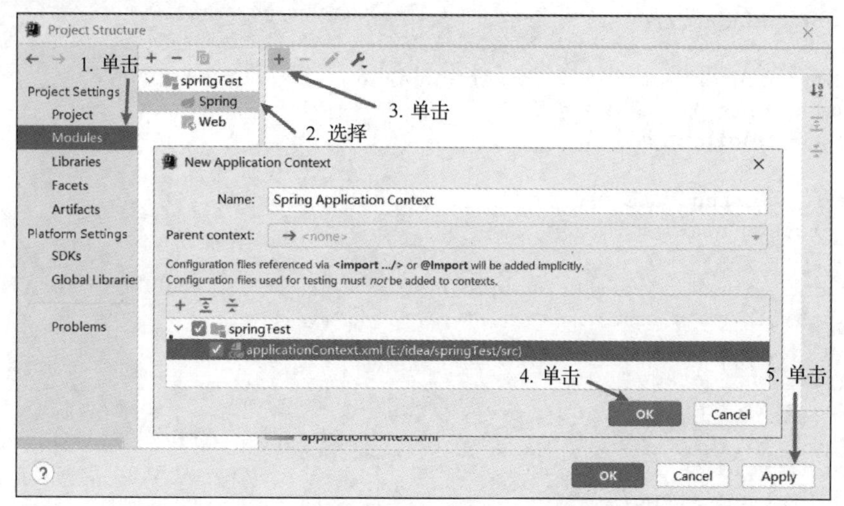

图 2.18 设置配置文件

下面编写配置文件，其中<bean>元素用于将一个对象注入容器，id 为此 Bean 对象的别名，此别名会在获取此 Bean 对象时用到，class 用于指定需要装配对象的路径，代码如例 2-5 所示。

【例 2-5】 applicationContext.xml

```
<?xml version="1.0" encoding="UTF-8"?>
<beans xmlns="http://www.springframework.org/schema/beans"
      xmlns:xsi="http://www.w3.org/2001/XMLSchema-instance"
      xsi:schemaLocation="http://www.springframework.org/schema/beans
          http://www.springframework.org/schema/beans/spring-beans.xsd">
   <!--id 为装入 Spring 容器时此对象的别名 class 为对象相对于 src 目录的路径 -->
   <bean id="student" class="Student"></bean>
</beans>
```

3．容器加载配置文件

在 src 目录下创建一个 Test 类，利用 ApplicationContext 容器的子类来加载配置文件，此配置文件的路径为相对于 src 目录的路径，代码如例 2-6 所示。

【例 2-6】 Test.java

```
public class Test {
   public static void main(String[] args) {
      ClassPathXmlApplicationContext classPathXmlApplicationContext =
           new ClassPathXmlApplicationContext("applicationContext.xml");
      Object student = classPathXmlApplicationContext.getBean("student");
      System.out.println(student);
   }
}
```

在例 2-6 的代码中，使用 getBean()方法获取 classPathXmlApplicationContext 容器中名为 student 的 Bean 对象，输出结果如图 2.19 所示。

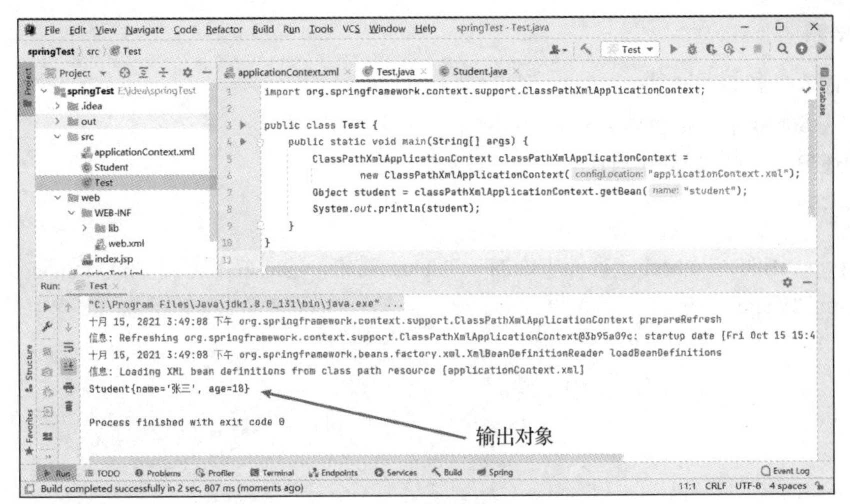

图 2.19 输出结果

2.4 本章小结

本章首先讲解了 Spring 的相关知识，接着通过示例讲解了 Bean 对象的创建与获取方法。通过对本章的学习，读者需要掌握向 Spring 容器中添加 Bean 对象的方法。

2.5 习题

1．填空题

（1）Spring 最核心的理念是________和________。

（2）Spring 容器是根据________思想来构建的。

2．选择题

（1）在 ApplicationContext 接口的实现类中，通过配置文件的路径加载容器信息的是（　　）。

A. ClassPathXmlApplicationContext　　B. FileSystemXmlApplicationContext

C. AnnotationConfigApplicationContext　　D. WebApplicationContext

（2）关于 Spring IoC 的特点，下列描述错误的是（　　）。

A. IoC 就是指程序之间的关系由程序代码直接操控

B. 控制反转是指控制权由应用代码转交到外部容器，即控制权的转移

C. IoC 将管理对象的职责搬进了框架中，并把它从应用代码中脱离出来

D. 在 Spring 配置文件中配置的 Bean 对象，将会在启动 Spring 容器时被加载到 Spring 容器

（3）关于 Bean 对象，下列说法正确的是（　　）。

A. Bean 对象由开发人员创建，随后交由 Spring 容器管理

B. 在整个业务流程中，涉及的所有对象都是 Bean 对象

C. 在容器的运行过程中，Spring 提供了向容器中添加 Bean 对象的方法

D. 在容器运行过程中，开发人员可以更改 Bean 对象的相关信息

3．思考题

（1）简述 IoC 和 DI 的概念。

（2）简述 Spring 容器的创建方式。

（3）简述 Bean 对象的创建与获取方法。

4．编程题

编写 Spring 用例，向 Spring 容器中添加一个 Dog 对象，随后取出容器中的 Dog 对象，Dog 对象的属性有 name 与 age。

第 3 章 Spring 中 Bean 的注入

本章学习目标

- 理解 Spring 中 Bean 的作用域。
- 理解 Bean 的生命周期。
- 掌握 Bean 的注入方式及复杂注入。
- 掌握利用注解管理 Bean 的方式。

Spring 中 Bean 的注入

Bean 的流转与赋值是 Spring 的核心。利用 IoC 将 Bean 交由容器管理可以简化开发，提升效率，因此 Bean 的基本操作是 Spring 基础知识中应重点掌握的部分。本章将讲述 Spring 中 Bean 的属性注入和利用注解管理 Bean 的方式，读者需要熟练掌握 Bean 的相关操作，了解 Bean 的作用域与生命周期。

3.1 Bean 的注入方式

Bean 的注入，即创建带有属性值的 Bean 对象，专业术语为 Bean 的实例化。Bean 的注入方式一般分为 3 种，分别为构造器注入、属性注入和接口注入，接口注入方式不要求掌握，本节将详细讲解构造器注入和属性注入。读者需要初步了解构造器注入，熟练掌握属性注入。

3.1.1 构造器注入

构造器注入指的是在被注入的类中声明一个构造方法，而构造方法可以是有参或无参的。Spring 在读取配置信息后，通过反射方式调用构造方法。如果是有参构造方法，可以向构造方法中传入所需的参数值，进而创建类对象。

在实际操作中，需要先创建一个类的构造方法，然后向配置文件中传入该类构造器的值，之后 Spring 将此类以配置文件中的参数进行实例化并将其加入容器中。

下面通过一个示例演示构造器注入。

1．创建被注入的类

创建 Dog 对象，如例 3-1 所示（在 IDEA 中通过 Alt+Ins 组合键可以生成 get()方法、set()方法、toString()方法和构造方法）。

【例 3-1】 Dog.java

```
1.  public class Dog {
2.      String name;
3.      int age;
4.
```

```
5.      public Dog() {
6.      }
7.      public Dog(String name, int age) {
8.          this.name = name;
9.          this.age = age;
10.     }
11.     public String getName() {
12.         return name;
13.     }
14.     public void setName(String name) {
15.         this.name = name;
16.     }
17.     public int getAge() {
18.         return age;
19.     }
20.     public void setAge(int age) {
21.         this.age = age;
22.     }
23.     @Override
24.     public String toString() {
25.         return "Dog{" +
26.                 "name='" + name + '\'' +
27.                 ", age=" + age +
28.                 '}';
29.     }
30. }
```

在以上代码中，Dog 类提供了所有属性的 get()方法、set()方法、toString()方法和有参构造方法。下面将 Dog 类写入配置文件，观察将其从容器中取出的结果。

2．配置文件的构造器注入

向 applicationContext.xml 文件中添加 Dog 类的配置，如例 3-2 所示。

【例 3-2】 applicationContext.xml

```
1.  <?xml version="1.0" encoding="UTF-8"?>
2.  <beans xmlns="http://www.springframework.org/schema/beans"
3.         xmlns:xsi="http://www.w3.org/2001/XMLSchema-instance"
4.         xsi:schemaLocation="http://www.springframework.org/schema/beans
5.             http://www.springframework.org/schema/beans/spring-beans.xsd">
6.      <!--name 为构造方法中的参数名，value 值为此参数的值-->
7.      <bean id="myDog" class="Dog">
8.          <constructor-arg name="age" value="3"/>
9.          <constructor-arg name="name" value="旺财"/>
10.     </bean>
11. </beans>
```

以上代码向<bean>元素中插入了构造方法元素<constructor-arg>，<bean>元素用于配置一个 Bean 对象，<constructor-arg>元素的作用是配置此 Bean 的构造参数。构造方法元素的 name 属性是构造方法中的参数名，value 属性是参数值。将此 Bean 命名为 myDog。值得一提的是，必须将所有的参数填写完毕才可以创建。如果少写了 name 参数或 age 参数，是无法创建此 Bean 的。

3．获取容器中的 Bean

在 Test 类中利用容器 ApplicationContext 的子类 ClassPathXmlApplicationContext 获取容器中注入的

myDog，查看其属性，如例 3-3 所示。

【例 3-3】 Test.java

```
1.  public class Test {
2.      public static void main(String[] args) {
3.          ApplicationContext applicationContext =
4.                new ClassPathXmlApplicationContext("applicationContext.xml");
5.          /*使用 getBean()方法获取容器中的 myDog 实例*/
6.          Object dog = applicationContext.getBean("myDog");
7.          System.out.println(dog);
8.      }
9.  }
```

以上代码通过 getBean()方法获取容器中的 myDog 实例，输出结果如图 3.1 所示。

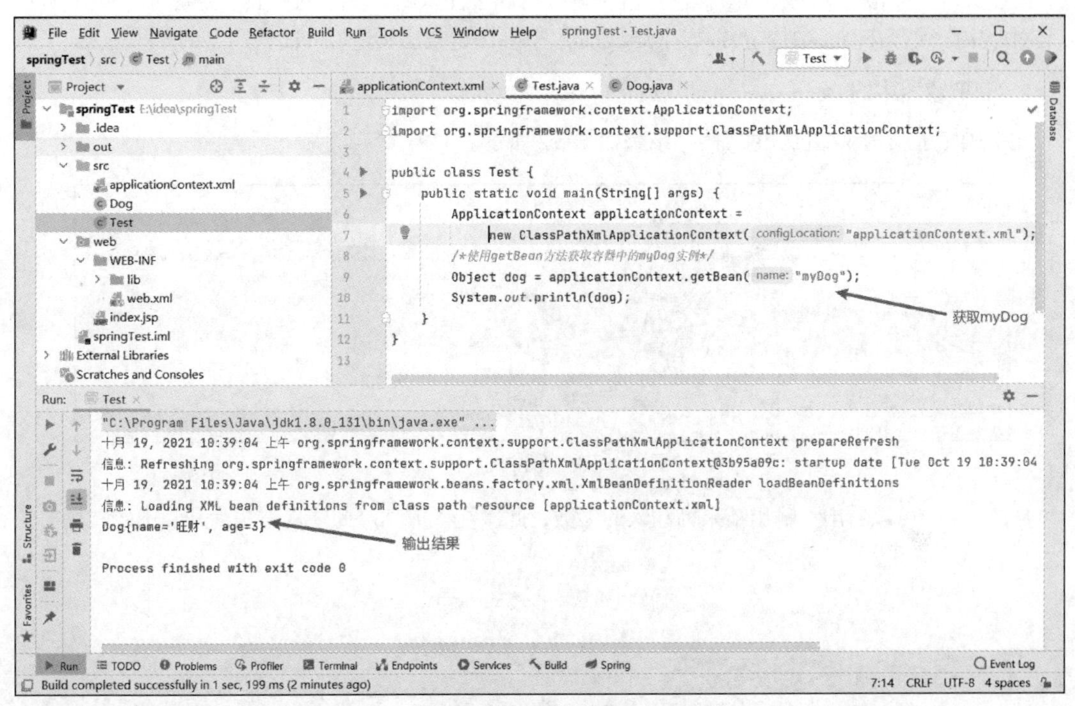

图 3.1　获取容器中的 myDog

从图 3.1 中可以看出，控制台输出了 name 为“旺财”的 Bean 对象，这说明通过构造器注入成功。此方法在开发中并不常用，读者了解即可。

3.1.2 属性注入

属性注入为开发中常用的注入方式，此方式通过 Bean 的 set()方法直接设置属性值，具体步骤与构造器注入类似。下面展示属性注入的示例，本示例采用例 3-1 的实体类作为注入对象。

1．配置文件的属性注入

配置文件的属性注入采用<property>元素，如例 3-4 所示。其中 name 表示属性的名称，value 表示此属性的值。

【例 3-4】 applicationContext.xml

```
1.  <?xml version="1.0" encoding="UTF-8"?>
```

```
2.  <beans xmlns="http://www.springframework.org/schema/beans"
3.         xmlns:xsi="http://www.w3.org/2001/XMLSchema-instance"
4.         xsi:schemaLocation="http://www.springframework.org/schema/beans
5.             http://www.springframework.org/schema/beans/spring-beans.xsd">
6.      <bean id="myDog" class="Dog">
7.          <!--property 元素用于对一个参数进行注入，name 为属性名，value 为此属性的值-->
8.          <property name="age" value="13"/>
9.          <property name="name" value="大旺财"/>
10.     </bean>
11. </beans>
```

值得一提的是，这种注入方式是通过调用对象中的 set()方法来实现注入的，如果 Dog 类没有 setName()方法，则 name 不能注入。属性注入与构造器注入的不同之处在于，构造器注入只需要 Dog 类有配置文件中对应参数的构造方法。配置文件进行属性注入时，传入的参数不需要对应构造方法，但是对象中必须存在空参数构造方法，Spring 通过空参构造方法构造后才能使用 set()方法注入。

2．获取容器中的 Bean

此次操作采用例 3-4 的示例代码，结果如图 3.2 所示。

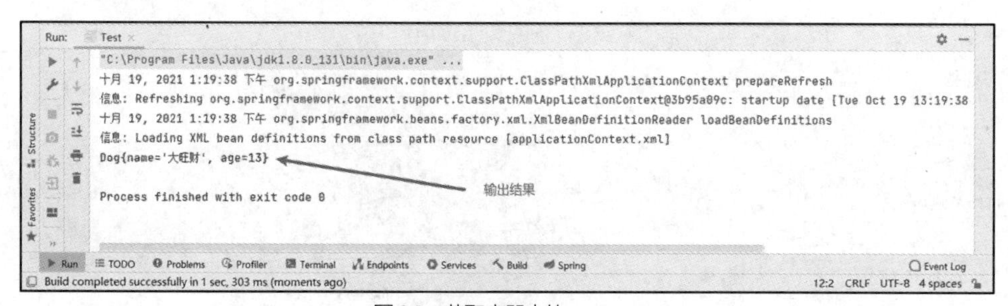

图 3.2　获取容器中的 myDog

从图 3.2 中可以看出，输出结果为注入的参数，此次注入成功。此注入方法为开发中的常用方法，读者需要熟练掌握。

3.2 Bean 的复杂注入

上一节主要介绍了 Bean 的两种注入方式，本节将针对属性注入的高级扩展进行讲解，内容为 Bean 的集合注入和对象注入，以及跨 Bean 取用属性值的方法。读者需要深入了解 Bean 的常用属性，初步掌握集合与对象的注入，初步了解 Bean 之间属性的传递。

3.2.1 Bean 的常用属性

在介绍复杂注入之前，先介绍 Bean 的详细配置。XML 文件中的<bean>元素有很多属性，在 3.1 节中提到了 id 和 class 属性，本小节把常用的属性进行列举，如表 3.1 所示。

表 3.1　Bean 的常用属性

属性	说明
class	指定 Bean 对象相对于 src 文件夹的路径
name	指定 Bean 对象的一个标识

续表

属性	说明
id	指定 Bean 对象的唯一标识，通常使用 id 从容器中取出 Bean 对象
scope	设置 Bean 对象的作用域
lazy-init	设置是否延迟加载，默认为 false，即非延迟加载
init-method	对象的初始化方法
destroy	对象的销毁方法

3.2.2 集合与对象的注入

当对象的属性为集合或对象类型时，注入方式和之前有所不同。下面详细讲解集合和对象的注入方式。

1．创建实例

创建实体类对象，代码分别如例 3-5、例 3-6 和例 3-7 所示（因 set()方法、get()方法、toString()方法和构造方法过多，在此全部忽略，实际操作中需要全部添加，两个构造方法分别为空参数构造方法和全参数构造方法）。

【例 3-5】 Teacher.java

```
1.  public class Teacher {
2.      MyClass myClass;
3.      List<String> position;
4.      List<Student> students;
5.      Map<String,Integer> score;
6.      String[] vacation;
7.  }
```

【例 3-6】 Student.java

```
1.  public class Student {
2.      int age;
3.      String name;
4.  }
```

【例 3-7】 MyClass.java

```
1.  public class MyClass {
2.      String name = "三班";
3.  }
```

在例 3-5、例 3-6 和例 3-7 中，例 3-5 为老师对象，其属性有学生集合和班级，例 3-6 与例 3-7 分别为学生和班级的实体类。下面对 Teacher 类进行注入。

2．编写配置文件

编写配置文件，将 Student 和 MyClass 注入 Teacher 类中，代码如例 3-8 所示。

【例 3-8】 applicationContext.xml

```
1.  <?xml version="1.0" encoding="UTF-8"?>
2.  <beans xmlns="http://www.springframework.org/schema/beans"
3.        xmlns:xsi="http://www.w3.org/2001/XMLSchema-instance"
4.        xsi:schemaLocation="http://www.springframework.org/schema/beans
5.            http://www.springframework.org/schema/beans/spring-beans.xsd">
```

```
6.     <!--学生1-->
7.     <bean id="student1" class="Student">
8.        <property name="age" value="13"/>
9.        <property name="name" value="张三"/>
10.    </bean>
11.    <!--学生2-->
12.    <bean id="student2" class="Student">
13.       <property name="age" value="18"/>
14.       <property name="name" value="李四"/>
15.    </bean>
16.    <!--班级-->
17.    <bean id="myClass" class="MyClass"/>
18.
19.    <!--老师-->
20.    <bean id="teacher" class="Teacher">
21.       <!--老师的班级-->
22.       <property name="myClass">
23.          <ref bean="myClass"></ref>
24.       </property>
25.       <!--老师的职位-->
26.       <property name="position">
27.          <list>
28.             <value>数学老师</value>
29.             <value>班主任</value>
30.          </list>
31.       </property>
32.       <!--老师的学生-->
33.       <property name="students">
34.          <list>
35.             <ref bean="student1"></ref>
36.             <ref bean="student2"></ref>
37.          </list>
38.       </property>
39.       <!--老师的教学评分-->
40.       <property name="score">
41.          <map>
42.             <entry key="数教" value="98"></entry>
43.             <entry key="英教" value="89"></entry>
44.          </map>
45.       </property>
46.       <!--老师的休假-->
47.       <property name="vacation">
48.          <array>
49.             <value>3天</value>
50.             <value>2天</value>
51.          </array>
52.       </property>
53.    </bean>
54. </beans>
```

在例 3-8 中，代码的第 6～15 行定义了两个学生，分别为张三和李四，第 17 行定义了一个班级，第 19～53 行定义了老师。学生 Bean 对象的构造不再讲解，下面讲解老师 Bean 对象的构造。

在老师的第一个属性中用<ref>元素的 bean 属性注入对象，其值为班级 Bean 对象的 id，<ref>元素的作用为规定一个引用对象，此对象可以给对象属性赋值。

老师的第二个属性为 list 集合，<list>元素是一个集合元素，在此采用<list>元素包裹<value>元素，通过一个<value>元素可添加一个值，只需加上多个<value>元素即可完成集合的注入。

老师的第三个属性为对象集合，采用<list>元素包裹<ref>元素完成注入。

老师的第四个属性为 map 集合，采用<map>元素加<entry>元素完成此 Bean 对象的注入，<entry>元素中的 key 和 value 分别代表 map 中的键和值。

老师的第五个属性为数组，采用<array>元素包裹<value>元素实现数组的注入。

集合对象的注入只需要初步理解，在项目中用到的频率不高。

3．获取容器中的 Bean

利用 getBean()方法获取 Teacher 对象，结果如图 3.3 所示。

```
"C:\Program Files\Java\jdk1.8.0_131\bin\java.exe" ...
十月 20, 2021 1:29:25 下午 org.springframework.context.support.ClassPathXmlApplicationContext prepare
信息: Refreshing org.springframework.context.support.ClassPathXmlApplicationContext@3b95a09c: startu
十月 20, 2021 1:29:25 下午 org.springframework.beans.factory.xml.XmlBeanDefinitionReader loadBeanDefi
信息: Loading XML bean definitions from class path resource [applicationContext.xml]
Teacher{myClass=MyClass@694e1548, position=[数学老师, 班主任], students=[Student{age=13, name='张三'},
```

图 3.3　获取容器中的 Teacher 对象

从图 3.3 中可以看出，Teacher 中的所有对象属性都输出了，配置文件中的值全部注入。值得一提的是，集合及对象的注入方法并不需要熟练掌握，当需要用到时再进行记忆更合适。

3.2.3 Bean 之间属性的传递

在配置文件中，可以使用表达式来获取相关的值，如例 3-9 所示。

【例 3-9】 applicationContext.xml

```
1.  <?xml version="1.0" encoding="UTF-8"?>
2.  <beans xmlns="http://www.springframework.org/schema/beans"
3.         xmlns:xsi="http://www.w3.org/2001/XMLSchema-instance"
4.         xsi:schemaLocation="http://www.springframework.org/schema/beans
5.             http://www.springframework.org/schema/beans/spring-beans.xsd">
6.      <!--学生1-->
7.      <bean id="student1" class="Student">
8.          <property name="age" value="13"/>
9.          <property name="name" value="张三"/>
10.     </bean>
11.     <!--学生2-->
12.     <bean id="student2" class="Student">
13.         <property name="age" value="#{student1.age}"/>
14.         <property name="name" value="李四"/>
15.     </bean>
16.
17. </beans>
```

从例 3-9 中可以看出，学生 2 的属性值 age 可以取用学生 1 的相关属性值，格式为#{id.properties}，结果是李四的年龄被注入为 13 岁，测试结果在此不做展示。

3.3 Bean 的作用域

当 Spring 容器创建一个 Bean 时，可以指定其作用域。Bean 的作用域指的是此 Bean 相对于其他 Bean 的可见范围。本节将对 Bean 的作用域进行介绍，读者需要掌握 Bean 常用的作用域。

3.3.1 作用域的种类

Spring 中 Bean 的作用域有 7 种，如表 3.2 所示。

表 3.2 Bean 的作用域

作用域名称	说明
singleton	指定 Bean 为单例，一个 Spring 容器中只有一个实例。无论获取多少次，返回的都是同一个对象
prototype	指定 Bean 为多例，每次通过 Spring 获取的对象都是新建的
request	对于每一次请求，容器都会返回一个新的实例
session	对于一次 session 请求，在客户端的 JSessionId 未失效的情况下，获取到的 Bean 都是同一个
globalSession	对于一次 globalSession 请求，请求到的 Bean 都是同一个，在 Web 环境中与 session 作用域相同，仅在 portlet 上下文有效
application	为每一个 ServletContext 对象分配一个新的 Bean
websocket	为每一个 websocket 对象分配一个新的 Bean

表 3.2 所示的 7 种作用域中，只有 singleton 和 prototype 作用域较常用，在一些不常见的情况下可能会用到 request、session、application 和 websocket 作用域。因此，下一小节将以 singleton 和 prototype 作用域为例讲解 Spring 作用域的作用。

3.3.2 singleton 与 prototype 作用域

singleton 是 Spring 默认的作用域，使用此作用域，获取相同 id 的 Bean，取得的都是相同的对象。此作用域广泛应用于无状态的 Bean，可减少资源消耗。例如，Spring 中常见的 Service 模块和 Mapper 模块均采用 singleton 作用域。

prototype 作用域为原型作用域，使用此作用域的 Bean，除第一次返回原地址外，以后从 Spring 容器中获取时都会以载入容器时的 Bean 为原型返回新的对象。

Bean 的作用域通过<bean>元素的 scope 指定，示例代码如下所示。

```
<bean id="myClass" class="MyClass" scope="singleton"/>
```

上述代码中的 scope 是作用域元素，可以指定表 3.2 中的 7 种作用域。以例 3-7 的 MyClass 对象进行测试（去掉 toString()方法，打印地址编码），将其<bean>元素的 scope 分别赋值为 singleton 和 prototype，两次测试结果分别如图 3.4 和图 3.5 所示。

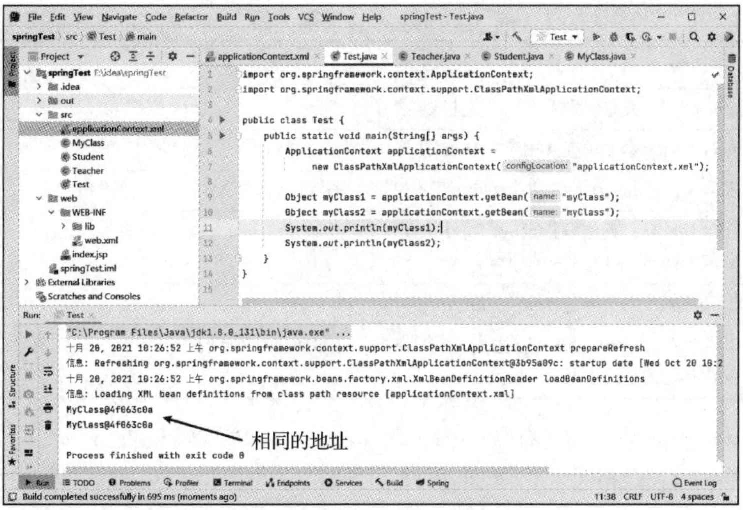

图3.4 singleton作用域测试

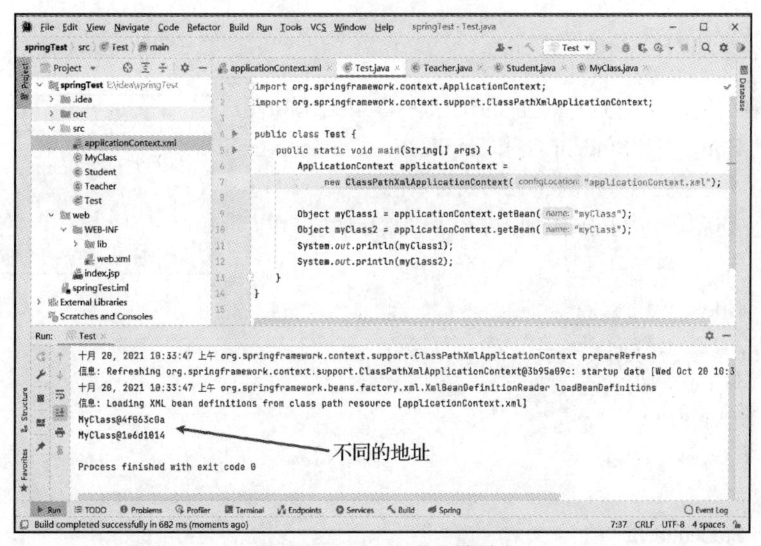

图3.5 prototype作用域测试

由图3.4和图3.5可以看出，通过prototype原型作用域每次获取的对象都不相同，而通过singleton单例作用域每次获取的对象均相同。

3.4 利用注解管理Bean

前面讲解了利用配置文件注入Bean的方法，这样得到的程序不宜整理和维护。本节讲解利用注解注入Bean的方式，开发人员只需编写对应的注解即可实现属性的注入。此外，本节还将介绍包扫描组件，它可以实现Bean的自动配置。读者必须熟练掌握本节的全部内容。

3.4.1 Bean的常用注解

Bean相关的注解非常多，在此介绍常用的几种，如表3.3所示。

表 3.3　Bean 的常用注解

注解名称	说明
@Component	类注解，被此注解标注的类将注入容器中
@Resource	在 Spring 容器中查找相应的 Bean 并将其注入属性中，优先选用名称相符合的 Bean 注入
@Autowired	与@Resource 注解的功能相同，只不过该注解优先选择容器中类型符合的 Bean 注入
@Value	属性注解，此注解指定的值将注入属性中，和@Resource 注解不同的是，此注解注入的是基本类型，@Resource 注解注入的是类类型
@Scope	类注解，用于指定 Bean 对象的作用域
@Qualifier	与@Autowired 注解一起使用，当容器中存在同类型的 Bean 时，@Autowired 注解会报错，需要通过此注解指定的名称注入

除了以上 6 种注解，常用的还有@Service、@Repository 和@Controller 注解，这 3 个注解的作用均与@Component 注解完全相同。不同的名称只是用来区分每个注解的用途，具体的用途将在第 10 章进行讲解。

3.4.2　注解的应用

下面以一个示例来讲解这些注解的用法。目录结构如图 3.6 所示，其中 Book 类、Student 类和 applicationContext.xml 文件分别如例 3-10、例 3-11 和例 3-12 所示，在此省略 set()、get()等方法，在实际操作中必须添加。Test 类为启动类，在此省略示例代码。

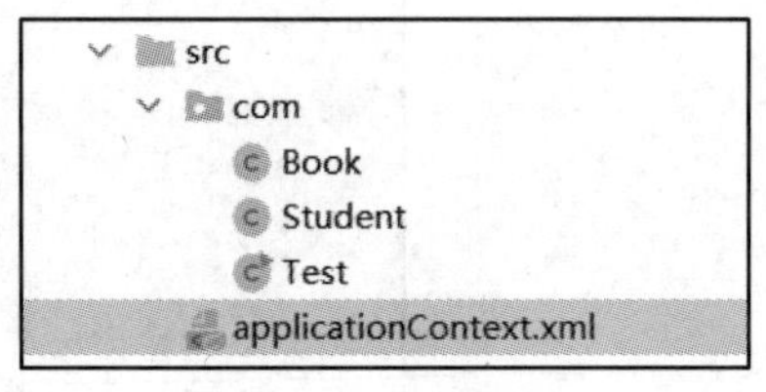

图 3.6　目录结构

【例 3-10】 Book.java

```
1.  @Component("myBook")
2.  public class Book {
3.      @Value("语文")
4.      String name;
5.  }
```

【例 3-11】 Student.java

```
1.  @Component(value = "myStudent")
2.  @Scope("prototype")
3.  public class Student {
4.
5.      @Value("12")
6.      int age;
7.
8.      @Value("王五")
9.      String name;
10.
11.     @Resource
12.     Book book;
13.
14.     @Qualifier("myBook")
15.     @Autowired
16.     Book book1;
17.
18. }
```

【例 3-12】 applicationContext.xml

```
1.  <?xml version="1.0" encoding="UTF-8"?>
2.  <beans xmlns="http://www.springframework.org/schema/beans"
3.        xmlns:xsi="http://www.w3.org/2001/XMLSchema-instance"
4.        xmlns:context="http://www.springframework.org/schema/context"
5.        xsi:schemaLocation="http://www.springframework.org/schema/beans
6.  http://www.springframework.org/schema/beans/spring-beans.xsd
7.  http://www.springframework.org/schema/context
8.  http://www.springframework.org/schema/context/spring-context.xsd">
9.      <!--包扫描注解-->
10.     <context:component-scan base-package="com"></context:component-scan>
11. </beans>
```

下面以例 3-11 中的 Student 类为例，详解介绍其中的注解。

（1）Student 类上方@Component 注解的作用为将此对象加入容器中，其后方括号中的 value 属性值为此 Bean 的名称。value 属性为默认属性，可以像 Book 类的@Component 注解一样省略。

（2）age 与 name 属性上方的是@Value 注解，此注解将其后方括号中的值注入属性中。

（3）在@Component 注解下方的是@Scope 注解，此注解规定了 Student 类在容器中的作用域，可选的作用域为表 3.2 中的 7 个作用域。

（4）Student 类中 book 属性的上方为@Resource 注解，此注解用于搜索容器中的 Bean 并将其注入属性中。搜索过程为先根据名称搜索，若容器中没有对应名称的 Bean，则改为类型搜索。@Resource 注解中有 name 属性，若指定此属性的值，则直接按照名称搜索。在此例中，先在容器中搜索名称为 book 的 Bean（Book 类在注入容器时，名称为 myBook），未能找到，因此采用类型搜索，搜寻容器中类型为 Book 的类，然后注入。

（5）Student 类中 book1 属性的上方为@Autowired 注解，此注解同样用于将容器中的 Book 类注入属性中。与@Resource 注解不同的是，此注解采用类型搜索，若无对应类型，则直接报错，如果有多个此类型的 Bean，同样报错。在有多个同类型 Bean 的情况下，需要通过@Qualifier 注解指定某种类型，此注解默认的 value 值为搜索的 Bean 名称，Spring 会根据此名称搜索 Bean 来注入。

注解配置完毕以后，需要让 Spring 扫描 com 包下的所有类，使这些注解生效。

配置 applicationContext.xml 文件，添加包扫描注解 context:component-scan，并将被扫描的包添加进即可，base-package 是相对于 src 目录的包路径，如例 3-12 所示。

3.5 Bean 的生命周期

本节主要讲解 Bean 的创建、初始化和销毁过程。在较复杂的情况下经常需要控制 Bean 的生命周期，即在 Bean 的初始化前后或销毁前后添加一些必要的操作。在 Spring 中，控制 Bean 的生命周期是一个很复杂的操作，不容易被初学者掌握。此节的目的是使读者初步了解 Bean 的生命周期，作为练习，读者可以利用 init()和 destroy()方法在 Bean 的初始化和销毁前后做一些简单的输出操作。

Bean 的生命周期如图 3.7 所示。

在图 3.7 中，背景为深色的模块为 Bean 的生命周期的主体，分别为 Bean 的实例化、Bean 的属性注入、Bean 的初始化和 Bean 的销毁。整个 Bean 生命周期的执行过程如下所示。

（1）执行 postProcessBeforeInstantiation()方法。此方法为 InstantiationAwareBeanPostProcessor 后置处理器的一个实现方法，其主要作用为返回代理类。若此类已经有代理类，经过此方法处理后将代理类返回，不再执行以下操作。若没有代理类，继续执行下面的操作创建 Bean。

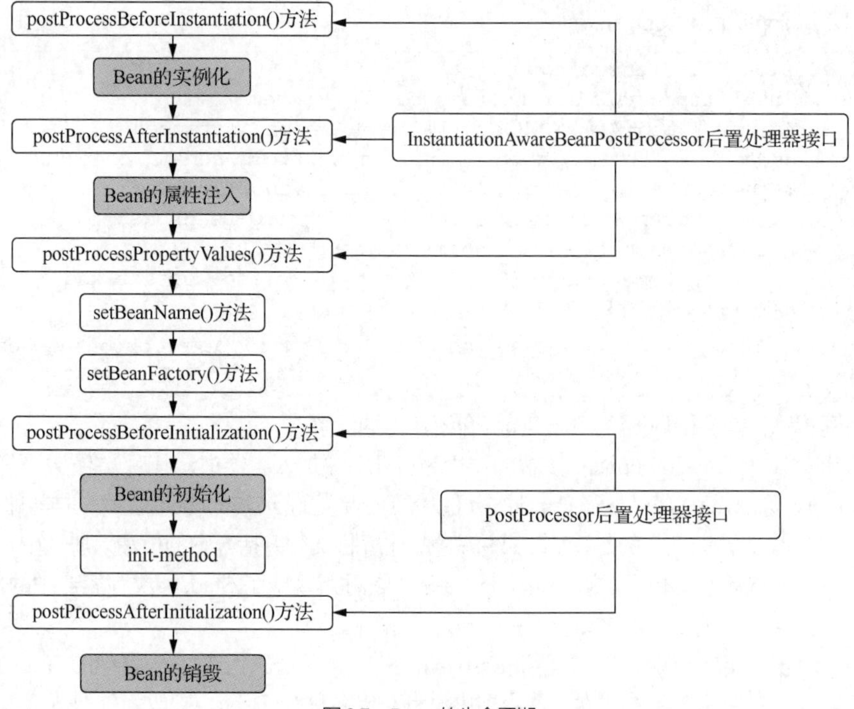

图 3.7　Bean 的生命周期

（2）进行 Bean 的实例化，即执行构造方法，此时类的所有参数都将被赋为默认值。

（3）在实例化 Bean 以后，执行 postProcessAfterInstantiation()方法。此方法在属性注入之前、实例化之后执行，其主要作用为对实例化后的 Bean 进行相关操作。若此方法返回 false，则终止后续步骤的执行。若返回 true，继续执行后续操作。

（4）执行 Bean 的属性注入操作，主要为基本的属性赋值操作，包括@Value 注解、@Autowired 注解和配置文件的设置等。

（5）postProcessPropertyValues()方法在属性注入之后执行，可以更改属性值。

（6）setBeanName()方法和 setBeanFactory()方法是为了方便在赋值之后进行相关操作的方法。两者的作用分别为设置类的属性和获取 Bean 工厂进行相关操作。

（7）postProcessBeforeInitialization()方法的功能为在初始化之前对属性进行更改。

（8）执行 Bean 的初始化操作，此操作需要对象实现 InitializationBean 接口，覆盖 afterPropertiesSet()方法，随后通过 afterPropertiesSet()方法进行相关操作。

（9）init-method 为开发人员指定的执行类，当生命周期执行到此时，执行此类中的 init 方法。

（10）postProcessAfterInitialization()方法在初始化之后执行，其作用为在初始化之后进行相关操作。

（11）Bean 的销毁发生在容器销毁时，这需要指定相关方法，在容器销毁时，指定的方法将会被调用。

下面举例说明 Bean 的初始化和销毁操作，例 3-13 与例 3-14 分别为注解的操作与配置文件的操作。

【例 3-13】 Student.java

```
1.  @Component(value = "myStudent")
2.  public class Student {
3.
4.      @PostConstruct
5.      public void init(){
6.          System.out.println("初始化");
```

```
7.      }
8.
9.      @PreDestroy
10.     public void destroy(){
11.         System.out.println("销毁");
12.     }
13. }
```

【例 3-14】 applicationContext.xml

```
1.  <?xml version="1.0" encoding="UTF-8"?>
2.  <beans xmlns="http://www.springframework.org/schema/beans"
3.        xmlns:xsi="http://www.w3.org/2001/XMLSchema-instance"
4.        xmlns:context="http://www.springframework.org/schema/context"
5.        xsi:schemaLocation="http://www.springframework.org/schema/beans
6.      http://www.springframework.org/schema/beans/spring-beans.xsd
7.      http://www.springframework.org/schema/context
8.      http://www.springframework.org/schema/context/spring-context.xsd">
9.    <bean id="student"
10.         class="com.Student"
11.         init-method="init"
12.         destroy-method="destroy">
13.
14.   </bean>
15.
16. </beans>
```

例3-13通过@PostConstruct注解与@PreDestroy注解来标注方法为Bean的初始化方法和销毁方法。例3-14通过init-method和destroy-method属性来指定方法为Bean的初始化方法和销毁方法。

3.6 本章小结

本章首先介绍了Bean的构造器注入和属性注入，接着详细介绍了复杂注入中的集合注入和对象注入，还讲解了Bean的作用域和常用注解，最后讲解的Bean的生命周期是Spring高阶内容需要深入学习的知识。有兴趣的读者可以提前了解Bean生命周期的相关内容，以便后续对Spring高阶内容的学习。

在本章内容中，属性注入相较于构造器注入更常用，读者需要着重掌握属性注入。另外，Bean的复杂注入与作用域在实际开发中不常用，读者只需初步了解。本章的重点是3.4节，读者需熟练掌握Bean的常用注解，这对后面内容的学习有很大的帮助。

3.7 习题

1．填空题

（1）向Spring容器中的Bean对象赋值的操作称为________。

（2）Spring中Bean的默认作用域是________。

（3）在Spring中，标记在类上后，可以将此类加入容器中的注解是________。

2．选择题

（1）关于<bean>元素的相关属性，下列描述错误的是（　　）。

A. class 属性用于指定 Bean 对象的绝对路径

B. scope 属性用于指定 Bean 的作用域

C. id 属性用于指定 Bean 对象在容器中的名称，此名称唯一

D. init-method 属性可以指定对象的初始化方法

（2）依赖注入的方式不包括（　　）。

A. 构造方法注入　B. set()方法注入　C. 接口注入　D. 注解注入

3．思考题

（1）简述 Bean 作用域的种类。

（2）简述属性注入的操作步骤。

（3）简述 Bean 的生命周期。

4．编程题

向 Spring 容器中添加 Dog 对象，并为其注入属性，属性为 name 和 age。要求使用两种方法实现，一种是使用配置文件属性方式注入，另一种是使用注解方式注入。

第4章 Spring 中的 AOP

本章学习目标

- 理解 AOP 的概念。
- 理解 AOP 的实现方式与实现原理。
- 掌握 AOP 的基本操作。
- 掌握 AOP 中切面优先级的配置方法。

Spring 中的 AOP

本章将讲解 Spring 进阶部分的 AOP。AOP 是 Spring 进阶内容中较复杂且重要的核心知识点，全称为 Aspect Oriented Programming，即面向切面编程，广泛应用于事务管理、日志管理、性能监控、权限检查等业务。本章将对 AOP 进行详细介绍。

4.1 AOP 简介

本节将讲解 AOP 的概念，读者需要初步掌握 AOP 的功能实现方法，熟悉 AOP 中的核心概念模块。

4.1.1 AOP 的基本概念

与 OOP 的面向对象思想类似，AOP 指的是抽取重复加在业务前后的代码，通过切面的方式应用在方法上。由于 AOP 的概念较抽象，下面以示例进行讲述，具体代码如例 4-1 和例 4-2 所示。

【例 4-1】 BookHandler.java

```
1.  package com;
2.  /*图书处理类*/
3.  public class BookHandler {
4.
5.      public void addBook(Book book){
6.          /*添加 book 操作*/
7.          System.out.println("添加 book");
8.      }
9.  }
```

【例 4-2】 StudentHandler.java

```
1.  package com;
2.  /*学生管理类*/
3.  public class StudentHandler {
```

```
4.
5.     public void addStudent(Student student){
6.         /*新增 Student 操作*/
7.         System.out.println("新增 student");
8.     }
9. }
```

例 4-1 和例 4-2 所示的是两个业务类，分别负责添加书和新增学生。现在需要在添加书和新增学生之前将操作记录到日志，并且在添加操作完成后，将添加结果也记录到日志里。

首先以非 AOP 的方式完成日志的记录，具体代码如例 4-3 和例 4-4 所示。

【例 4-3】 BookHandler.java

```
1. package com;
2. /*图书处理类*/
3. @Component
4. public class BookHandler {
5.
6.     public void addBook(Book book){
7.         /*添加日志*/
8.         System.out.println("开始添加");
9.
10.        try {
11.            /*添加 book 操作*/
12.            System.out.println("添加 book");
13.        }catch (Exception e){
14.            /*添加失败日志*/
15.            System.out.println("添加失败");
16.        }
17.        /*添加成功日志*/
18.        System.out.println("添加成功");
19.    }
20.
21. }
```

【例 4-4】 StudentHandler.java

```
1. package com;
2. /*学生管理类*/
3. @Component
4. public class StudentHandler {
5.
6.     public void addStudent(Student student){
7.         /*添加日志*/
8.         System.out.println("开始添加");
9.
10.        try {
11.            /*新增 student 操作*/
12.            System.out.println("添加 student");
13.        }catch (Exception e){
14.            /*添加失败日志*/
15.            System.out.println("添加失败");
```

```
16.         }
17.         /*添加成功日志*/
18.         System.out.println("添加成功");
19.     }
20. }
```

在例 4-3 和例 4-4 中，加粗的代码为业务代码，此业务代码在实际开发中较复杂，在业务代码前后加入代码管理日志，会使代码很臃肿。除此之外，在原有的业务类中附加代码，侵入性过强，且不易维护。在实际开发中，需要输出日志的业务模块很多，这种日志管理方式的复用性较差，因此，一般不采用此方式处理日志。

下面演示利用 AOP 方式实现日志管理。首先导入 aspectj.weaver.jar 和 aopalliance.jar 包作为 AOP 的支持，然后编写日志处理业务代码，代码如例 4-5 所示（实体类仍采用例 4-1 和例 4-2 的代码）。

【例 4-5】 LogAopAspect.java

```
1.  /*日志处理切面类*/
2.  @Component
3.  @Aspect
4.  public class LogAopAspect {
5.      /*在切点之前执行*/
6.      @Before("execution(* com.*Handler.add*(..))")
7.      public void dealAddLog(JoinPoint joinPoint){
8.          /*记录日志*/
9.          System.out.println("开始添加");
10.     }
11.     /*在切点抛出异常时执行*/
12.     @AfterThrowing("execution(* com.*Handler.add*(..))")
13.     public void dealExceptionLog(JoinPoint joinPoint){
14.         /*记录日志*/
15.         System.out.println("添加失败");
16.     }
17.     /*在切点返回之前执行*/
18.     @AfterReturning("execution(* com.*Handler.add*(..))")
19.     public void dealAfterReturnAddLog(JoinPoint joinPoint){
20.         /*记录日志*/
21.         System.out.println("添加成功");
22.     }
23. }
```

在例 4-5 所示的代码中，在类的上方添加@Aspect 注解，标明此类为切面处理类。此后，在此类的方法上标注的通知注解就可以生效。

dealAddLog()方法上方的@Before 注解是一个通知注解，添加此注解的方法会在目标方法执行之前执行。目标方法通常指通知注解中 execution 表达式指定的方法。常用的通知注解将在 4.2.1 小节讲述。

在例 4-1 中，dealAddLog()方法上方的@Before 通知注解对应的目标方法为 execution 表达式指定的方法。execution 表达式较复杂，将在 4.2.2 小节做详细介绍。在此指出，上述代码中的目标方法为 addBook()方法和例 4-2 中的 addStudent()方法。

同理，用@AfterThrowing 注解标注的方法会在目标方法抛出异常后执行，用@AfterReturning 注解标注的方法会在目标方法返回值后执行。

除此之外，还需要配置 applicationContext.xml 文件，如例 4-6 所示。

【例 4-6】 applicationContext.xml

```
1.  <?xml version="1.0" encoding="UTF-8"?>
2.  <beans xmlns="http://www.springframework.org/schema/beans"
3.          xmlns:xsi="http://www.w3.org/2001/XMLSchema-instance"
4.          xmlns:context="http://www.springframework.org/schema/context"
5.          xmlns:aop="http://www.springframework.org/schema/aop"
6.          xsi:schemaLocation="http://www.springframework.org/schema/beans
7.     http://www.springframework.org/schema/beans/spring-beans.xsd
8.     http://www.springframework.org/schema/context
9.     http://www.springframework.org/schema/context/spring-context.xsd
10.    http://www.springframework.org/schema/aop
11.    http://www.springframework.org/schema/aop/spring-aop-4.3.xsd">
12.
13. <context:component-scan base-package="com"></context:component-scan>
14.     <aop:aspectj-autoproxy/>
15. </beans>
```

例 4-6 中加粗代码的作用为声明 AOP 注解，只有加上此声明，才能使 AOP 相关的注解生效。配置 Test 类的代码如例 4-7 所示。

【例 4-7】 applicationContext.xml

```
1.  package com;
2.
3.  public class Test {
4.      public static void main(String[] args) {
5.          ApplicationContext applicationContext =
6.                new ClassPathXmlApplicationContext("applicationContext.xml");
7.
8.          BookHandler bookHandler =
9.             (BookHandler)applicationContext.getBean("bookHandler");
10.         StudentHandler studentHandler =
11.            (StudentHandler)applicationContext.getBean("studentHandler");
12.
13.         bookHandler.addBook(new Book());
14.         studentHandler.addStudent(new Student());
15.     }
16. }
```

上述代码的输出结果如图 4.1 所示。

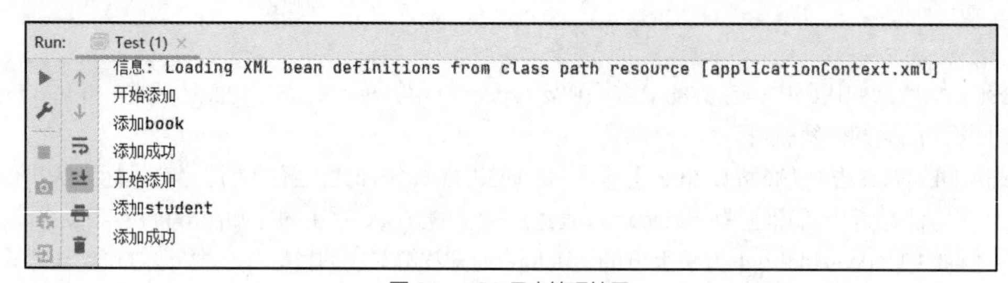

图 4.1　AOP 日志管理结果

在上面的示例中，通过 AOP 配置一个切面类来切入指定的方法，在此方法的前后执行相关操作。这种方式没有对目标方法进行更改，相较于日志的传统管理方式，此方式不仅可以降低代码的侵入性，而且可以提高代码的复用性。

4.1.2 AOP 的核心概念

上小节提到了目标方法的概念，目标方法其实就是需要处理的方法。在例 4-1 中，addBook()方法就是一个目标方法。在 AOP 中，类似目标方法的概念还有很多，下面详细介绍一些 AOP 的核心概念。

1．连接点（JoinPoint）

连接点是一个非常抽象的概念，它表示一个时机。例如，某个类的初始化完成后、某个方法执行前或程序出现异常时，都是连接点。但在 Spring AOP 中，仅仅支持方法的连接点。在例 4-1 中，addBook()方法执行前、执行后和发生异常时均为连接点。

2．通知（Advice）

通知是指在连接点处执行的某段代码，例 4-5 的 dealAddLog()方法就是一个通知。除此之外，所有被通知注解标注的方法均为通知。常用的通知注解将在 4.2.1 小节讲述。

3．切点（PointCut）

切点是匹配连接点的一个抽象概念。在例 4-5 中，execution(* com.*Handler.add*(..))表达式是一个切点，此切点匹配了 BookHandler 类中的“addBook()方法执行前”连接点和 StudentHandler 类中的“addStudent()方法执行前”连接点。切点表达式如何匹配连接点是 AOP 实现的核心要点。

4．目标对象（Target）

目标对象是指需要处理的目标业务类，例 4-1 中的 BookHandler 类可以作为目标对象。若没有 AOP 的支持，目标对象就需要独立完成所有的业务逻辑。目标方法通常指切点映射的方法。

5．切面（Aspect）

切面指 AOP 处理的类，此类封装了许多需要插入其他类执行的代码。例 4-5 中的 LogAopAspect 类就是切面类。

6．织入（Weaving）

织入是指将通知添加到目标类具体连接点的过程。将例 4-5 中的 dealAddLog()方法添加到例 4-1 中“addStudent()方法执行前”连接点的过程就是织入。

4.2 Spring 中 AOP 的实现方式

本节将主要讲解 Spring 中 AOP 的实现方式，分别为基于注解方式和基于 XML 方式。读者需要熟练掌握基于注解方式实现 AOP，初步了解基于 XML 方式实现 AOP。

4.2.1 基于注解实现 AOP

基于注解实现 AOP 为开发中常用的方式。为了方便开发，AOP 提供了一系列的注解，常用注解如表 4.1 所示。

表 4.1 常用注解

注解名称	说明
@Aspect	此注解用于标注一个类为一个切面，供容器读取
@Pointcut	此注解用在方法上，用于指定一个切点

续表

注解名称	说明
@Before	指定一个前置通知，此通知将在目标方法之前执行
@After	指定一个后置通知，此通知将在目标方法之后执行
@Around	指定一个环绕通知，相当于同时配置了前置通知与后置通知
@AfterReturning	指定一个返回通知，此通知将在目标方法返回值之后执行
@AfterThrowing	指定一个异常通知，此通知将在目标方法抛出异常时执行

下面通过注解的方式编写一个切面类，目标对象选用例 4-1 中的 BookHandler 类，此切面类如例 4-8 所示。

【例 4-8】 LogAopAspect.java

```
1.  /*日志处理切面类*/
2.  @Component
3.  @Aspect
4.  public class LogAopAspect {
5.      @Pointcut("execution(* com.BookHandler.addBook(..))")
6.      public void pointCut(){}
7.
8.      /*在目标方法之前执行*/
9.      @Before("pointCut()")
10.     public void dealAddBeforeLog(JoinPoint joinPoint){
11.         System.out.println("开始添加");
12.     }
13.     /*在目标方法之后执行*/
14.     @After("pointCut()")
15.     public void dealAddAfterLog(JoinPoint joinPoint){
16.         System.out.println("添加结束");
17.     }
18.     @Around("pointCut()")
19.     public void dealAddAroundLog(ProceedingJoinPoint pjb){
20.         System.out.println("执行Around环绕 添加之前");
21.         try {
22.             /*执行目标方法*/
23.             pjb.proceed();
24.         } catch (Throwable e) {
25.             System.out.println("发现异常");
26.         }
27.         System.out.println("执行Around环绕 处理完成");
28.     }
29.     @AfterReturning("pointCut()")
30.     public void dealAfterReturnAddLog(JoinPoint joinPoint){
31.         System.out.println("方法返回成功");
32.     }
33.     /*在目标方法抛出异常时执行*/
34.     @AfterThrowing("execution(* com.BookHandler.addBook(..))")
35.     public void dealExceptionLog(JoinPoint joinPoint){
36.         System.out.println("添加失败");
37.     }
38. }
```

下面详细介绍例 4-8 中的所有注解。

（1）通过@Component 注解将 LogAopAspect 类加入 Spring 容器中，然后在 LogAopAspect 类上方标注@Aspect 注解，被@Aspect 注解标注的类将识别或切面类，之后即可在类中编写通知的相关代码。

（2）用@Pointcut 注解标注的方法为切点，切点需要用 execution 表达式指定。此表达式将在 4.2.2 小节介绍。

（3）用@Before 注解标注的方法为前置通知，此通知将在指定的目标方法之前执行。

（4）@After 注解与@Before 注解类似，不同的是，用@After 注解标注的通知在指定的目标方法之后执行，而用@Before 注解标注的通知在目标方法之前执行。

（5）用@Around 注解标注的方法为环绕通知。环绕通知的参数为 ProceedingJoinPoint 类，此类代表一个执行连接点，调用此类的 proceed()方法即可执行目标方法，在调用此类的 proceed()方法之前的所有代码将在目标方法之前执行。其效果等同于用@After 注解标注的前置通知，在调用 proceed()方法之后的代码将在目标方法之后执行。

（6）@AfterReturning 注解与@After 注解类似，不同的是，用@AfterReturning 注解标注的通知在目标方法返回值之后执行，而用@After 注解标注的通知在目标方法之后执行。当目标方法内有异常发生时，用@AfterReturning 注解标注的通知将不会执行，而用@After 注解标注的通知正常执行。

（7）用@AfterThrowing 注解标注的通知只会在目标对象抛出异常时执行。

配置 applicationContext.xml 文件来启用 AOP 注解，配置文件的代码如例 4-6 所示。下面通过测试类测试输出结果，测试类的代码如例 4-9 所示。

【例 4-9】 Test.java

```
1.  import org.springframework.context.ApplicationContext;
2.  import org.springframework.context.support.*;
3.
4.  public class Test {
5.      public static void main(String[] args) {
6.          ApplicationContext applicationContext =
7.              new ClassPathXmlApplicationContext("applicationContext.xml");
8.
9.          BookHandler bookHandler =
10.             (BookHandler)applicationContext.getBean("bookHandler");
11.         StudentHandler studentHandler =
12.             (StudentHandler)applicationContext.getBean("studentHandler");
13.         bookHandler.addBook(new Book());
14.     }
15. }
```

执行例 4-9 所示的代码，结果如图 4.2 所示。

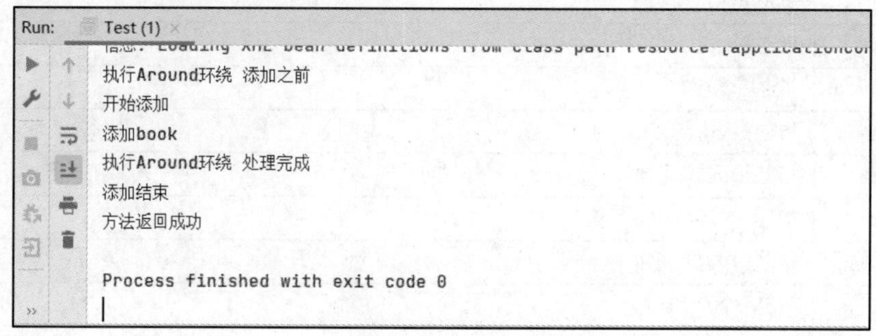

图 4.2 测试结果

由图 4.2 可以看出，执行顺序为@Around 注解中 Proceed()方法之前的通知、@Before 注解的前置通知、目标方法、@Around 注解中 Proceed()方法之后的通知、@After 注解的后置通知和@AfterReturning 注解的返回通知。

当目标方法出现异常时，执行将中断于目标方法的位置，随后执行@AfterThrowing 注解的异常通知。

4.2.2 execution 表达式

在 4.2.1 小节讲解了基于注解实现 AOP 的示例，示例中使用 execution 表达式来配置切点到目标方法的映射。下面讲解开发中常见的 execution 表达式的使用。

execution 表达式的一般格式如下所示。

```
execution(<修饰符> <返回类型> <包名>.<类名>.<方法名>(<参数>))
```

以例 4-1 为例，将 addBook()方法作为目标方法，execution 表达式的完整写法如下所示。

```
execution(public void com.BookHandler.addBook(Book))
```

其中 public 修饰符、包名、类名可以省略，其他不可以省略。在开发中，通常只省略修饰符，写法如下所示。

```
execution(void com.BookHandler.addBook(Book))
```

当需要指定多个方法时，可以根据方法名的特征选用“*”作为参数。将例 4-1 和例 4-2 中的 addBook()和 addStudent()方法作为目标方法，execution 表达式的格式如下所示。

```
execution(* com.*Handler.add*(**))
```

在以上代码中，修饰符省略；返回值类型标注为“*”，表示所有返回值类型均匹配；包名设置为包的全名；类名为“*Handler”，表示匹配任何名称以“Handler”结尾的类；方法名为“add*”，表示匹配所有名称以“add”开头的方法；参数为“**”，表示匹配任意个、任意类型的参数。

4.2.3 基于 XML 实现 AOP

下面讲解利用 XML 实现 AOP，这种方式在开发中很少使用，读者只需初步了解。Spring 提供了一系列的 XML 元素用于实现 AOP，具体如表 4.2 所示。

表 4.2　XML 元素

元素名称	说明
<aop:config>	AOP 实现的根元素
<aop:aspect>	指定一个切面类
<aop:advisor>	指定一个通知
<aop:pointcut>	指定一个切点
<aop:before>	指定一个前置通知
<aop:around>	指定一个环绕通知
<aop:after-returning>	指定一个返回通知
<aop:after-throwing>	指定一个异常通知

4.2.1 小节讲解了常用注解的用法，XML 实现与注解实现的核心思想大致相同，但 XML 实现步骤

烦琐，代码量大，且不易于维护，因此开发中大多使用注解实现。接下来以示例演示基于 XML 实现 AOP。

以例 4-1 中的 BookHandler 类为目标对象，创建切面类，切面类如例 4-10 所示。

【例 4-10】 LogAopAspectXML.java

```
1.  package com;
2.
3.  import org.aspectj.lang.JoinPoint;
4.  import org.aspectj.lang.ProceedingJoinPoint;
5.  import org.aspectj.lang.annotation.*;
6.  import org.springframework.stereotype.Component;
7.
8.  /*日志处理切面类*/
9.  @Component
10. public class LogAopAspectXML {
11.
12.     /*在目标方法之前执行*/
13.     public void dealAddBeforeLog(){
14.        System.out.println("开始添加");
15.     }
16.     /*在目标方法之后执行*/
17.     public void dealAddAfterLog(){
18.        System.out.println("添加结束");
19.     }
20.
21.     public void dealAddAroundLog(ProceedingJoinPoint pjb){
22.        System.out.println("执行 Around 环绕添加之前");
23.        try {
24.           /*执行目标方法*/
25.           pjb.proceed();
26.        } catch (Throwable e) {
27.           System.out.println("发现异常");
28.           return;
29.        }
30.        System.out.println("执行 Around 环绕 处理完成");
31.     }
32.     /*在目标方法返回值时执行*/
33.     public void dealAfterReturnAddLog(JoinPoint joinPoint){
34.        /*记录日志*/
35.        System.out.println("方法返回成功");
36.     }
37.
38.     /*在目标方法抛出异常时执行*/
39.     public void dealExceptionLog(JoinPoint joinPoint){
40.        /*记录日志*/
41.        System.out.println("添加失败");
42.     }
43. }
```

在例 4-10 中，使用@Component 注解将 LogAopAspectXML 类加入容器中，此类只配置了简单的方法，没有附加其他注解，接下来需要编写 applicationContext.xml 文件，将此类配置为切面类，

applicationContext.xml 文件如例 4-11 所示。

【例 4-11】 applicationContext.xml

```
1.  <?xml version="1.0" encoding="UTF-8"?>
2.  <beans xmlns="http://www.springframework.org/schema/beans"
3.         xmlns:xsi="http://www.w3.org/2001/XMLSchema-instance"
4.         xmlns:context="http://www.springframework.org/schema/context"
5.         xmlns:aop="http://www.springframework.org/schema/aop"
6.         xsi:schemaLocation="http://www.springframework.org/schema/beans
7.           http://www.springframework.org/schema/beans/spring-beans.xsd
8.           http://www.springframework.org/schema/context
9.           http://www.springframework.org/schema/context/spring-context.xsd
10.          http://www.springframework.org/schema/aop
11.          http://www.springframework.org/schema/aop/spring-aop-4.3.xsd">
12.     <!--扫描 com 包下的所有注解-->
13.     <context:component-scan base-package="com"></context:component-scan>
14.     <!--AOP 配置-->
15.     <aop:config>
16.         <!--指定切点-->
17.         <aop:pointcut
18.                 id="pointcut"
19.                 expression="execution(* com.BookHandler.addBook(..))"
20.         />
21.         <!--指定切面-->
22.         <aop:aspect ref="logAopAspectXML">
23.             <!--指定前置通知-->
24.             <aop:before method="dealAddBeforeLog" pointcut-ref="pointcut"/>
25.             <!--指定后置通知-->
26.             <aop:after method="dealAddAfterLog" pointcut-ref="pointcut"/>
27.             <!--指定环绕通知-->
28.             <aop:around method="dealAddAroundLog" pointcut-ref="pointcut"/>
29.             <!--指定返回通知-->
30.             <aop:after-returning
31.                     method="dealAfterReturnAddLog"
32.                     pointcut-ref="pointcut"
33.             />
34.             <!--指定异常通知-->
35.             <aop:after-throwing
36.                     method="dealExceptionLog"
37.                     pointcut-ref="pointcut"
38.             />
39.         </aop:aspect>
40.
41.     </aop:config>
42. </beans>
```

下面解释例 4-11 中的元素。

（1）通过<aop:config>元素配置 AOP，此元素用于标明其中的代码为 AOP 配置。

（2）<aop:pointcut>元素用于标注切点，此元素需要用 expression 表达式来配置。

（3）<aop:aspect>元素用于指定切面类，需要用 ref 属性来指定一个切面类。在本例中，将例 4-10 中的切面类编写入此文件。需要注意的是，例 4-10 中的类已经被加入容器中，在此不需要写<bean>元素。

（4）<aop:before>元素用于指定前置通知，需要配置通知与切点。

（5）其他的通知注解与前置通知类似。

4.3 AOP 中切面的优先级

默认配置下，当多个切面类中的通知通过切点切入同一个目标方法时，执行顺序默认为切面类的名字的字典顺序。而在有些情况下需要按照特定的顺序来执行各个切点，此时就需要控制切面类的执行顺序。本节通过两种方式讲解如何控制切面的优先级。通过对本节的学习，读者需要熟练掌握基于注解和 Ordered 实现优先级配置的方式，初步了解基于 XML 实现优先级配置的方式。

4.3.1 基于注解和 Ordered 接口配置优先级

通过注解控制切面的优先级是开发中常用的方法。以例 4-1 中 BookHandler 类的 addBook()方法为目标方法，编写切面类，如例 4-12 和例 4-13 所示。

【例 4-12】 Aspect1.java

```
1.  /*日志处理切面类*/
2.  @Component
3.  @Aspect
4.  public class Aspect1 {
5.     /*在目标方法之前执行*/
6.      @Before("execution(* com.BookHandler.addBook(..))")
7.      public void addBefore(JoinPoint joinPoint){
8.          /*记录日志*/
9.          System.out.println("Aspect1 类执行 addBefore()方法");
10.     }
11.
12.     /*在目标方法之后执行*/
13.     @After("execution(* com.BookHandler.addBook(..))")
14.     public void addAfter(JoinPoint joinPoint){
15.         /*记录日志*/
16.         System.out.println("Aspect1 类执行 addAfter()方法");
17.     }
18. }
```

【例 4-13】 Aspect2.java

```
1.  /*日志处理切面类*/
2.  @Component
3.  @Aspect
4.  public class Aspect2 {
5.       /*在目标方法之前执行*/
6.      @Before("execution(* com.BookHandler.addBook(..))")
7.      public void addBefore(JoinPoint joinPoint){
8.          /*记录日志*/
9.          System.out.println("Aspect2 类执行 addBefore()方法");
10.     }
```

```
11.
12.     /*在目标方法之后执行*/
13.     @After("execution(* com.BookHandler.addBook(..))")
14.     public void addAfter(JoinPoint joinPoint){
15.         /*记录日志*/
16.         System.out.println("Aspect2 类执行 addAfter()方法");
17.     }
18. }
```

例 4-12 和例 4-13 均切入同一个目标方法，因 Aspect1 的字典顺序大于 Aspect2，所以执行顺序：Aspect1→Aspect2→目标方法。随后配置 application.xml 文件，添加包扫描并启用 AOP 注解（见例 4-6）。最后编写测试类，调用 addBook()方法，在此省略示例代码。最终运行结果如图 4.3 所示。

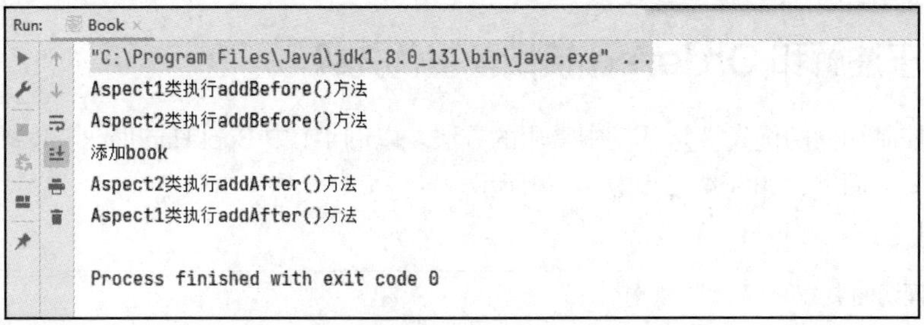

```
Run:  Book
"C:\Program Files\Java\jdk1.8.0_131\bin\java.exe" ...
Aspect1类执行addBefore()方法
Aspect2类执行addBefore()方法
添加book
Aspect2类执行addAfter()方法
Aspect1类执行addAfter()方法

Process finished with exit code 0
```

图 4.3　运行结果

下面使用@Order 注解来配置各个切面的执行顺序，如例 4-14 和例 4-15 所示。

【例 4-14】 Aspect1.java

```
1.  /*日志处理切面类*/
2.  @Component
3.  @Aspect
4.  @Order(1)
5.  public class Aspect1 {
6.      /*在目标方法之前执行*/
7.      @Before("execution(* com.BookHandler.addBook(..))")
8.      public void addBefore(JoinPoint joinPoint){
9.          /*记录日志*/
10.         System.out.println("Aspect1 类执行 addBefore()方法");
11.     }
12.
13.     /*在目标方法之后执行*/
14.     @After("execution(* com.BookHandler.addBook(..))")
15.     public void addAfter(JoinPoint joinPoint){
16.         /*记录日志*/
17.         System.out.println("Aspect1 类执行 addAfter()方法");
18.     }
19. }
```

【例 4-15】 Aspect2.java

```
1.  /*日志处理切面类*/
2.  @Component
3.  @Aspect
4.  @Order(0)
```

```
5.  public class Aspect2 {
6.      /*在目标方法之前执行*/
7.      @Before("execution(* com.BookHandler.addBook(..))")
8.      public void addBefore(JoinPoint joinPoint){
9.          /*记录日志*/
10.         System.out.println("Aspect2 类执行 addBefore()方法");
11.     }
12.
13.     /*在目标方法之后执行*/
14.     @After("execution(* com.BookHandler.addBook(..))")
15.     public void addAfter(JoinPoint joinPoint){
16.         /*记录日志*/
17.         System.out.println("Aspect2 类执行 addAfter()方法");
18.     }
19. }
```

在例 4-14 和例 4-15 中，@Order 注解标注在类上方，用来配置切面的优先级，括号中的数值越小，切面的优先级就越大。当配置此注解后，切面的执行顺序：Aspect2→Aspect1→目标方法。此处省略运行结果。

除此之外，还可以通过实现 Ordered 接口来实现切面优先级的配置，如例 4-16 和例 4-17 所示。

【例 4-16】 Aspect1.java

```
1.  /*日志处理切面类*/
2.  @Component
3.  @Aspect
4.  public class Aspect1 implements Ordered{
5.      /*在目标方法之前执行*/
6.      @Before("execution(* com.BookHandler.addBook(..))")
7.      public void addBefore(JoinPoint joinPoint){
8.          /*记录日志*/
9.          System.out.println("Aspect1 类执行 addBefore()方法");
10.     }
11.
12.     /*在目标方法之后执行*/
13.     @After("execution(* com.BookHandler.addBook(..))")
14.     public void addAfter(JoinPoint joinPoint){
15.         /*记录日志*/
16.         System.out.println("Aspect1 类执行 addAafter()方法");
17.     }
18.
19.     @Override
20.     public int getOrder() {
21.         return 1;
22.     }
23. }
```

【例 4-17】 Aspect2.java

```
1.  /*日志处理切面类*/
2.  @Component
3.  @Aspect
4.  public class Aspect2 implements Ordered{
```

```
5.     /*在目标方法之前执行*/
6.     @Before("execution(* com.BookHandler.addBook(..))")
7.     public void addBefore(JoinPoint joinPoint){
8.         /*记录日志*/
9.         System.out.println("Aspect2 类执行 addBefore()方法");
10.    }
11.
12.    /*在目标方法之后执行*/
13.    @After("execution(* com.BookHandler.addBook(..))")
14.    public void addAfter(JoinPoint joinPoint){
15.        /*记录日志*/
16.        System.out.println("Aspect2 类执行 addAfter()方法");
17.    }
18.
19.    @Override
20.    public int getOrder() {
21.        return 0;
22.    }
23. }
```

在例 4-16 和例 4-17 中，两个类均实现了 Ordered 接口，覆盖了 getOrder()方法，返回相应的优先级数值。与@Order 注解相同，返回的数值越小，切面的优先级越大，此处省略运行结果。

4.3.2 基于 XML 配置优先级

下面通过 XML 配置切面的优先级，此方式在开发中不常用。切面类代码如例 4-12 和例 4-13 所示，applicationContext.xml 文件的内容如例 4-18 所示。

【例 4-18】 applicationContext.xml

```
1. <!--扫描 com 包下的所有注解-->
2.     <context:component-scan base-package="com"></context:component-scan>
3.
4.     <aop:config>
5.         <!--切点-->
6.         <aop:pointcut
7.              id="pointCut"
8.              expression="execution(* com.BookHandler.addBook(..))"
9.         />
10.        <aop:aspect ref="aspect1" order="1">
11.            <aop:before method="addBefore" pointcut-ref="pointCut"/>
12.            <aop:after method="addAfter" pointcut-ref="pointCut"/>
13.        </aop:aspect>
14.        <aop:aspect ref="aspect2" order="0">
15.            <aop:before method="addBefore" pointcut-ref="pointCut"/>
16.            <aop:after method="addAfter" pointcut-ref="pointCut"/>
17.        </aop:aspect>
18.    </aop:config>
```

只需在<aop:aspect>元素中添加 order 属性即可配置切面的优先级，order 的数值越小，切面的优先级越大。

4.4 AOP 的实现原理

AOP 在 Spring 中是通过代理实现的。在 AOP 编程中，如果一个类被标记为目标对象，则 Spring 的 IoC 容器将根据此目标对象产生一个代理对象。在获取此目标对象时，返回的是代理对象。当该代理对象的方法被调用时，会依次执行目标方法的前置通知、目标方法和目标方法的后置通知。通过此方式，Spring 可以在目标方法前后进行相应的操作。本节将详细介绍代理设计模式和动态代理，读者需要深入理解代理设计模式的思想，初步了解动态代理的编写方式。

4.4.1 代理设计模式

Spring 的 IoC 容器通过动态代理来产生代理对象。在了解动态代理之前，先要了解代理设计模式。代理设计模式为 Java 面向对象设计模式中的一种，它的 UML（Unified Modeling Language，统一建模语言）类图如图 4.4 所示。

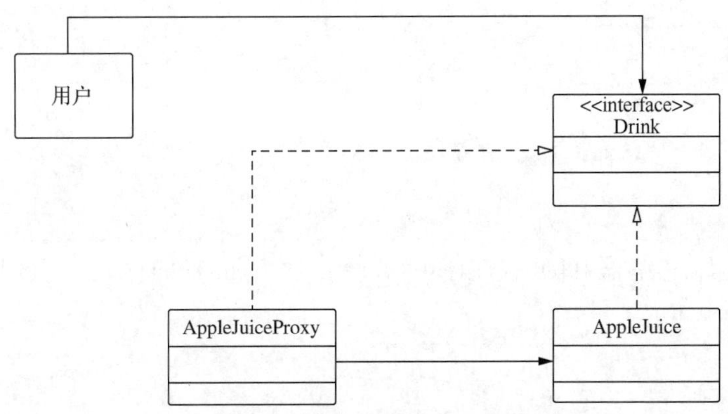

图 4.4　代理设计模式的 UML 类图

由图 4.4 可知，用户使用 Drink 类型的 AppleJuice 类进行相关操作，在此基础上对 AppleJuice 类进行扩展。使用代理设计模式：新建 AppleJuiceProxy 代理类来关联 AppleJuice 类并实现 Drink 接口，在 AppleJuiceProxy 代理类中覆盖 Drink 接口中的方法，并在此方法中调用 AppleJuice 类中对应的方法。

下面讲解代理设计模式的实现。

（1）创建 Drink 接口，在其中添加 myDrink()方法供子类实现，如例 4-19 所示。

【例 4-19】 Drink .java

```
1.  public interface Drink {
2.      /*喝饮料*/
3.      void myDrink();
4.  }
```

（2）编写目标对象 AppleJuice 类，使其实现 Drink 接口，覆盖其中的方法，如例 4-20 所示。

【例 4-20】 AppleJuice.java

```
1.  @Component
2.  public class AppleJuice implements Drink {
3.
4.      @Override
5.      public void myDrink() {
```

```
6.          System.out.println("喝饮料");
7.      }
8.  }
```

（3）编写代理对象 AppleJuiceProxy 类，使其实现 Drink 接口，覆盖其中的方法，同时添加目标对象属性。在 myDrink()方法中调用目标方法中的 myDrink()方法，并在其前后加上输出操作，如例 4-21 所示。

【例 4-21】 AppleJuiceProxy.java

```
1.  @Component
2.  public class AppleJuiceProxy implements Drink{
3.
4.      @Resource
5.      AppleJuice appleJuice;
6.
7.      @Override
8.      public void myDrink() {
9.
10.         System.out.println("喝饮料之前");
11.
12.         appleJuice.myDrink();
13.
14.         System.out.println("喝饮料之后");
15.     }
16. }
```

（4）创建测试类，从容器中取出代理对象，并调用其中的 myDrink()方法，如例 4-22 所示。

【例 4-22】 Test.java

```
1.  public class Test {
2.      public static void main(String[] args) {
3.          ApplicationContext applicationContext =
4.              new ClassPathXmlApplicationContext("applicationContext.xml");
5.          /*获取代理对象*/
6.          Drink appleJuiceProxy =
7.              (Drink)applicationContext.getBean("appleJuiceProxy");
8.
9.          appleJuiceProxy.myDrink();
10.     }
11. }
```

运行例 4-22 的代码，输出结果如图 4.5 所示。

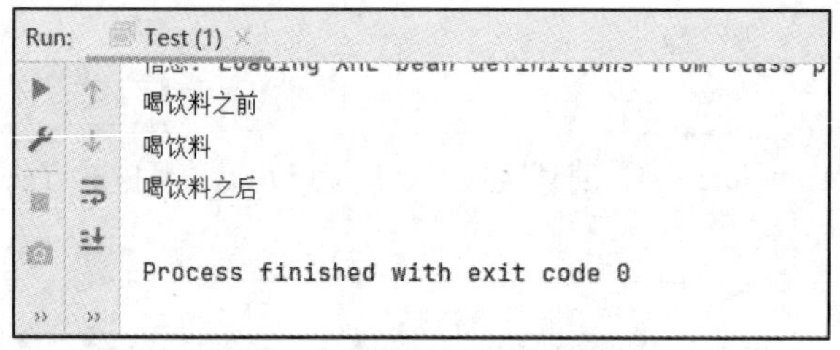

图 4.5　代理设计模式的测试结果

以上示例将目标对象隐藏，用户只能接触代理对象，调用代理对象的方法，然后由代理对象的方法调用目标对象的目标方法，完成代理。

4.4.2 JDK 动态代理

4.4.1 小节的代理方式有缺陷，即只能由开发人员手动定义代理方法。在实际开发中，需要编写的代理对象很多，无法全部交给开发人员编写。下面介绍 JDK 动态代理。以例 4-19 和例 4-20 为基础，创建测试类，在测试类中通过 JDK 动态代理创建代理对象，如例 4-23 所示。

【例 4-23】 Test.java

```
1.  public class Test {
2.      public static void main(String[] args) {
3.          ApplicationContext applicationContext =
4.              new ClassPathXmlApplicationContext("applicationContext.xml");
5.
6.          /*获取目标对象*/
7.          Drink appleJuice =
8.                  (Drink)applicationContext.getBean("appleJuice");
9.
10.         /*通过 Proxy 类的 newProxyInstance()方法直接创建代理对象*/
11.         Drink appleJuiceProxy = (Drink)Proxy.newProxyInstance(
12.                 Test.class.getClassLoader(),
13.                 AppleJuice.class.getInterfaces(),
14.                 new InvocationHandler() {
15.                     /*此方法将在代理对象的方法被调用时调用*/
16.                     @Override
17.                     public Object invoke(
18.                             Object proxy,
19.                             Method method,
20.                             Object[] args
21.                     ) throws Throwable {
22.                         System.out.println("喝饮料之前--JDK 动态代理");
23.                         /*调用目标对象的目标方法，参数 appleJuice 为目标对象*/
24.                         Object invoke = method.invoke(appleJuice, args);
25.                         System.out.println("喝饮料之后--JDK 动态代理");
26.                         return invoke;
27.                     }
28.                 }
29.         );
30.
31.         /*调用通过 JDK 动态代理创建的代理对象*/
32.         appleJuiceProxy.myDrink();
33.     }
34. }
```

从例 4-23 中可以看到，通过 Proxy 类的 newProxyInstance()方法可以直接创建代理对象。下面详细讲解 newProxyInstance()方法。

（1）newProxyInstance()方法的第一个参数为类构造器，JDK 动态代理需要指定一个类构造器来构造代理对象。此参数一般为应用类加载器，获取应用类加载器的方式有很多，在此使用类字节码获取应用类加载器。

（2）newProxyInstance()方法的第二个参数为被代理类的接口的字节码数组，可以通过被代理类的字节码获得。

（3）newProxyInstance()方法的第三个参数为 InvocationHandler 的实现类，在此使用匿名内部类构造此类，并覆盖 invoke()方法。invoke()方法将在代理对象的方法被调用时调用，其第一个参数为代理对象，第二个参数为被代理类的方法，第三个参数为调用方法的参数。在 invoke()方法被调用时，可以使用 method.invoke()方法反射调用目标对象的目标方法，从而实现在目标方法前后执行相关操作。

运行例 4-23 中的测试类，结果如图 4.6 所示。

```
Run:  Test (1)
信息: Loading XML bean definitions from class path resource [applicationContext.xm
喝饮料之前--JDK动态代理
喝饮料
喝饮料之后--JDK动态代理

Process finished with exit code 0
```

图 4.6　JDK 动态代理测试结果

4.4.3 CGLib 动态代理

动态代理除 JDK 提供的方式之外还有 CGLib 提供的方式，此方式称为 CGLib 动态代理。此代理方式不是通过接口，而是通过继承子类的方式产生代理对象。以例 4-19 和例 4-20 为基础，编写测试类，在测试类中通过 CGLib 动态代理创建代理对象，如例 4-24 所示。

【例 4-24】 Test.java

```
1.  public class Test {
2.      public static void main(String[] args) {
3.          ApplicationContext applicationContext =
4.              new ClassPathXmlApplicationContext("applicationContext.xml");
5.
6.          /*获取目标对象*/
7.          Drink appleJuice =
8.                  (Drink)applicationContext.getBean("appleJuice");
9.
10.         Enhancer enhancer = new Enhancer();
11.         enhancer.setSuperclass(AppleJuice.class);
12.         enhancer.setCallback(new MethodInterceptor() {
13.             @Override
14.             public Object intercept(
15.                     Object o,
16.                     Method method,
17.                     Object[] objects,
18.                     MethodProxy methodProxy) throws Throwable {
19.
20.                 System.out.println("喝饮料之前--CGLib 动态代理");
21.                 Object invoke = method.invoke(appleJuice, objects);
22.                 System.out.println("喝饮料之后--CGLib 动态代理");
23.
24.                 return invoke;
25.             }
26.         });
27.         AppleJuice appleJuiceProxy = (AppleJuice)enhancer.create();
28.
29.         /*调用通过 CGLib 动态代理创建的代理对象*/
30.         appleJuiceProxy.myDrink();
31.     }
32. }
```

从例 4-24 中可以看到，CGLib 动态代理需要设置父类对象与回调函数，之后通过 create()方法创建代理对象。下面详细讲解 CGLib 动态代理创建代理对象的方法。

（1）Enhancer 类的 setSuperclass()方法负责设置目标对象，setCallback()方法负责处理回调操作，并传入需要 MethodInterceptor 接口的实现类。

（2）MethodInterceptor 接口的实现类需要覆盖 intercept()方法，该方法的第一个参数为代理对象，第二个参数为目标方法，第三个参数为被代理方法的参数，第四个参数为代理方法。此方法的使用方法与例 4-23 中的 invoke()方法类似，通过 method 属性反射调用目标对象的目标方法即可。

运行测试类，结果如图 4.7 所示。

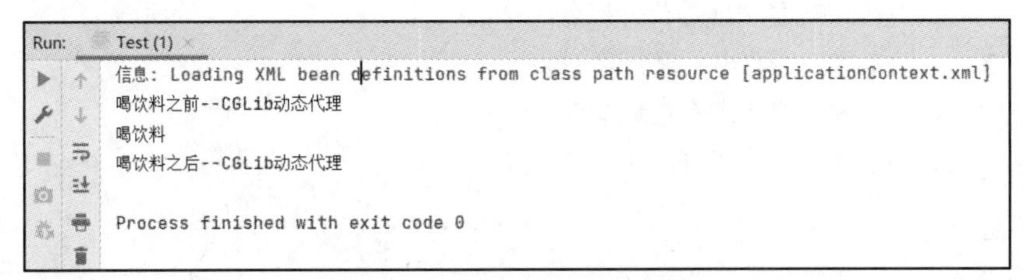

图 4.7　CGLib 动态代理测试结果

需要注意的是，CGLib 动态代理创建代理对象的方式不受接口的限制，但是不能代理 final 类型的方法；JDK 动态代理创建代理对象的方式受接口的限制，如果被代理类没有接口，将无法使用 JDK 动态代理。在 Spring 中，默认使用 JDK 动态代理，如果被代理类没有接口，将使用 CGLib 动态代理。

4.5 本章小结

本章首先介绍了 AOP 的概念，然后分别通过注解和 XML 方式实现了 AOP，接着讲解了 AOP 切面优先级的配置方法，最后讲解了 AOP 的实现原理。

在本章内容中，AOP 的概念和基于注解实现 AOP 是重点内容。基于 XML 实现 AOP 的方式在开发中很少使用。AOP 切面优先级的配置方法在复杂的开发中可能会用到，属于基本内容。AOP 的实现原理是面试中经常提到的知识点，主要考查 JDK 动态代理和 CGLib 动态代理的区别，属于需要深入了解的知识点，读者可按需了解。

4.6 习题

1．填空题

（1）动态代理的两种方式分别为________动态代理和________动态代理。

（2）可以配置 AOP 切面优先级的注解是________。

（3）Spring 中标记在某类上方后，可以将此类解析为切面类的注解是________。

2．选择题

（1）关于 AOP 的核心概念，下列描述错误的是（　　）。

A. 切点存在于切面类中，每个切面表达式都代表一个切点

B. 连接点的代码称为通知

C. 一个切点映射的所有连接点组成的面称为切面

D. 目标对象是指切点映射的连接点所在的类

（2）关于 AOP 的注解，下列说法错误的是（ ）。

A. 当目标方法内出现异常时，@AfterReturning 注解不会生效

B. @Pointcut 注解标注在某类上方时，表示此类为切面类

C. 在 AOP 的通知注解中，可以使用 execution 表达式指定多个连接点

D. @AfterThrowing 注解仅在目标方法出现异常时执行

（3）关于 AOP 代理，下列说法错误的是（ ）。

A. JDK 动态代理需要实现被代理对象的接口

B. CGLib 动态代理需要继承被代理对象

C. JDK 动态代理只能代理有接口的对象

D. CGLib 动态代理不能代理有接口的对象

3．思考题

（1）简述 AOP 的基本概念。

（2）简述使用注解实现 AOP 的方式。

（3）简述 JDK 动态代理和 CGLib 动态代理的区别。

4．编程题

创建一个 Dog 类，向其中加入 eat()方法。使用 AOP 在 eat()方法的前后输出任意一段文字（使用注解方式实现）。

第5章 Spring 与数据库的交互

本章学习目标

- 理解 Spring JDBC 的概念。
- 掌握使用 JdbcTemplate 操作数据库的方法。
- 掌握 JdbcTemplate 在日常开发中的使用。

Spring 与数据库的交互

问题是时代的声音，回答并指导解决问题是理论的根本任务。传统的 JDBC 代码烦琐，表关系维护困难，不利于日常开发。为了解决此类问题，Spring 提供了 Spring JDBC 来更加高效地操作数据库。本章作为 Spring 事务的过渡章，将简单讲解 Spring 与数据库的交互。

5.1 Spring JDBC 基础

Spring JDBC 是 Spring 的封装模块，它负责为数据库连接提供支持。本节将详细介绍 Spring JDBC 的配置。

5.1.1 Spring JDBC 简介

Spring JDBC 由 4 个部分组成，分别为 core（核心包）、object（对象包）、datasource（数据源包）和 support（支持包），其中较为核心的是 core 包中的 JdbcTemplate 类与 datasource 包中的 DriverManagerDataSource 类。

DriverManagerDataSource 类负责配置相关的数据库连接参数，其中核心参数为连接驱动、数据库地址、用户名和密码。配置完成后，通过 DriverManagerDataSource 类即可连接数据库进行数据操作。

JdbcTemplate 类封装了许多数据库之间的操作，属于操作数据库的工具类。工具类的使用需要实际的连接类来支撑，因此，JdbcTemplate 类需要配置 DriverManagerDataSource 类来连接数据库，从而进行数据库之间的数据交互。

5.1.2 Spring JDBC 的配置

1．下载 JDBC 依赖包

登录 Maven 官网，访问 JDBC 依赖包的下载地址，此处下载的是 mysql-connector-java-8.0.16.jar，与其兼容的 MySQL 数据库版本为 8.0 或 8.0 以上。

将下载完成的 JAR 包移至 lib 文件夹，右键单击此 JAR 包，在弹出的快捷菜单中选择 Add as Library 命令将其加入框架，具体操作如图 5.1 所示。

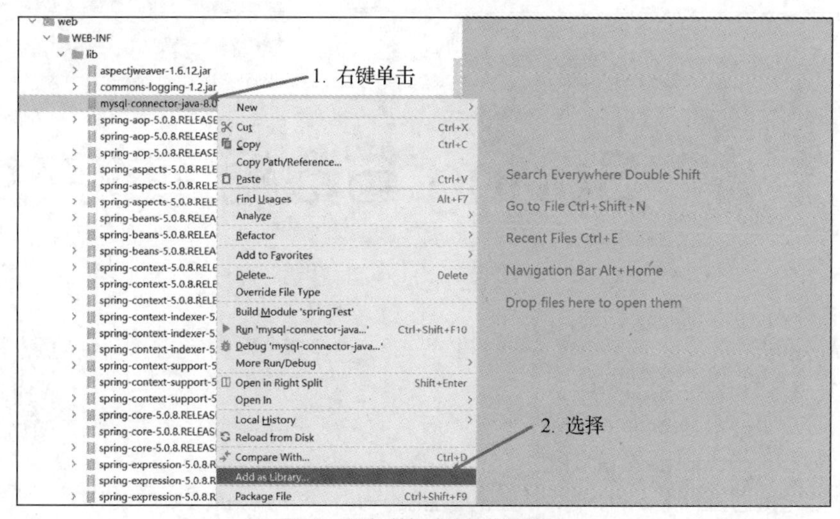

图 5.1　添加数据库连接的依赖包

2．配置 DriverManagerDataSource 类与 JdbcTemplate 类

以本地的 test 数据库为例，配置数据库的连接。applicationContext.xml 文件如例 5-1 所示。

【例 5-1】 applicationContext.xml

```
1.  <!--扫描 com 包下的所有注解-->
2.  <context:component-scan base-package="com"/>
3.
4.  <!--配置连接-->
5.  <bean id="myDataSource"
6.       class="org.springframework.jdbc.datasource.DriverManagerDataSource">
7.     <!--连接驱动-->
8.     <property  name="driverClassName" value="com.mysql.cj.jdbc.Driver"/>
9.     <!--连接地址-->
10.    <property
11.       name="url"
12.       value="jdbc:mysql://127.0.0.1/test?characterEncoding=utf8&amp;
13.              useSSL=false&amp;
14.              serverTimezone=UTC&amp;
15.              allowPublicKeyRetrieval=true"/>
16.    <!--用户名-->
17.    <property name="username" value="root" />
18.    <!--密码-->
19.    <property name="password" value="***"/>
20. </bean>
21.
22. <!--配置模板-->
23. <bean id="jdbcTemplate"
24.       class="org.springframework.jdbc.core.JdbcTemplate">
25.    <!--配置数据源-->
26.    <property name="dataSource" ref="myDataSource"/>
27. </bean>
```

下面讲解例 5-1 中的配置。

（1）以上代码的第 4~20 行配置 DriverManagerDataSource 类，并将其命名为 myDataSource。此类

需要4个参数，第一个参数为driverClassName，表示此次连接需要的驱动；第二个参数为url，表示此次要连接数据库的地址；第三个参数为username，表示连接数据库的用户名；第四个参数为password，表示此次连接数据库需要的密码。

（2）以上代码的第22～27行配置JdbcTemplate类，此类只需要DriverManagerDataSource类作为参数，参数名为dataSource。在之后的代码中将使用JdbcTemplate类进行数据库操作。

5.2 使用JdbcTemplate操作数据库

数据库交互的常用SQL语句可以分为3种：第一种为对表数据进行查询使用的语句，通常将其称为数据查询语言（Data Query Language，DQL）；第二种为对表数据进行增删改操作使用的语句，通常将其称为数据操纵语言（Data Manipulation Language，DML）；第三种为对表结构进行更改使用的语句，通常将其称为数据定义语言（Data Definition Language，DDL）。下面讲解利用JdbcTemplate完成以上提及的操作。

5.2.1 创建数据表

创建学生表stu，如图5.2所示。

id	name	course	score
1	张三	数学	90
2	张三	语文	50
3	张三	地理	40
4	李四	语文	55
5	李四	政治	45
6	王五	政治	30

图5.2 学生表stu

创建表stu的SQL语句如例5-2所示。

【例5-2】 stu.sql

```
1.  CREATE TABLE 'stu'  (
2.    'id' int(0) NOT NULL AUTO_INCREMENT,
3.    'name' varchar(255) CHARACTER,
4.    'course' varchar(255),
5.    'score' int(0),
6.   )
7.  INSERT INTO 'stu' VALUES (1, '张三', '数学', 90);
8.  INSERT INTO 'stu' VALUES (2, '张三', '语文', 50);
9.  INSERT INTO 'stu' VALUES (3, '张三', '地理', 40);
10. INSERT INTO 'stu' VALUES (4, '李四', '语文', 55);
11. INSERT INTO 'stu' VALUES (5, '李四', '政治', 45);
12. INSERT INTO 'stu' VALUES (6, '王五', '政治', 30);
```

5.2.2 DQL操作

1．单行单列返回结果的查询

从容器中取出jdbcTemplate对象，查询张三的数学成绩，如例5-3所示。

【例5-3】 Test.java

```
1.  public class Test {
2.      public static void main(String[] args) {
3.          ApplicationContext applicationContext =
4.               new ClassPathXmlApplicationContext("applicationContext.xml");
5.
6.          /*获取jdbcTemplate对象*/
```

```
7.          JdbcTemplate jdbcTemplate =
8.                  (JdbcTemplate)applicationContext.getBean("jdbcTemplate");
9.          /*查询 id 为 1 的学生的分数*/
10.         Integer integer =
11.                 jdbcTemplate.queryForObject(
12.                         "select score from stu where id=1",
13.                         Integer.class);
14.
15.         System.out.println(integer);
16.     }
17. }
```

在例 5-3 中，利用 JdbcTemplate 类的 queryForObject()方法可以对数据库中的数据进行查询操作。若返回结果为单行单列，此方法的第二个参数为基本数据类型包装类或 String 类型。因为此次查询结果为 int 类型，所以可以使用 Integer 包装类映射查询结果，输出结果如下所示。

```
90
```

2．单行多列返回结果的查询

在实际开发中，更多的是查询单行多列返回结果，可以通过 RowMapper 接口完成查询。此接口中定义了一个 mapRow()方法，该方法用于完成字段值和类属性值的映射（默认情况下数据库字段与属性名一一对应）。除此之外，Spring 为了开发方便，提供了 RowMapper 接口的实现类：BeanPropertyRowMapper 类。下面讲解查询的示例，如例 5-4 所示。

【例 5-4】 Test.java

```
1.  public class Test {
2.      public static void main(String[] args) {
3.          ApplicationContext applicationContext =
4.              new ClassPathXmlApplicationContext("applicationContext.xml");
5.
6.          /*获取 jdbcTemplate 对象*/
7.          JdbcTemplate jdbcTemplate =
8.                  (JdbcTemplate)applicationContext.getBean("jdbcTemplate");
9.          /*查询 id 为 1 的学生的课程信息*/
10.         Student student =
11.                 jdbcTemplate.queryForObject(
12.                         "select * from stu where id=1",
13.                         new BeanPropertyRowMapper<>(Student.class));
14.
15.         System.out.println(student);
16.     }
17. }
```

例 5-4 中的 Student 实体类如例 5-5 所示。在实际编写时，Student 类中需要添加 toString()方法、set()方法和 get()方法。

【例 5-5】 Student.java

```
1.  public class Student {
2.      String name;
3.      String course;
4.      Integer score;
5.  }
```

在例 5-4 中，queryForObject()方法的第二个参数为 BeanPropertyRowMapper 类，此类的构造方法

需要类字节码作为参数。当查询结果为单行多列时，结果将映射到此参数对应的类中。例 5-4 的输出结果如图 5.3 所示。

```
Stu{name='张三', course='数学', score=90}

Process finished with exit code 0
```

图 5.3　输出结果

在例 5.4 中，返回结果为 Student 类。需要注意的是，数据库数据到实体类数据的映射是通过 set()方法完成的，如果 Student 类中没有 set()方法，则无法完成映射。

3．多行多列返回结果的查询

当需要查询多条数据时，queryForObject()方法不再适用，可以使用 query()方法。具体查询代码如例 5-6 所示。

【例 5-6】 Test.java

```
1.  public class Test {
2.      public static void main(String[] args) {
3.          ApplicationContext applicationContext =
4.              new ClassPathXmlApplicationContext("applicationContext.xml");
5.
6.          /*获取 jdbcTemplate 对象*/
7.          JdbcTemplate jdbcTemplate =
8.                  (JdbcTemplate)applicationContext.getBean("jdbcTemplate");
9.          /*查询所有学生的信息*/
10.         List<Student> student =
11.                 jdbcTemplate.query(
12.                         "select * from stu where id=1",
13.                         new BeanPropertyRowMapper<>(Student.class));
14.
15.         System.out.println(student);
16.     }
17. }
```

例 5-6 中的 query()方法负责查询多条数据，其中的参数与 queryForObject()方法类似。

5.2.3 DML 操作

DML 操作包括对数据库中数据进行的增加、删除和修改操作，除去 SQL 语句的不同，这 3 种操作没有很大的区别。下面讲解利用 JdbcTemplate 类完成 DML 操作的方法，如例 5-7 所示。

【例 5-7】 Test.java

```
1.  public class Test {
2.      public static void main(String[] args) {
3.          ApplicationContext applicationContext =
4.              new ClassPathXmlApplicationContext("applicationContext.xml");
5.
6.          /*获取 jdbcTemplate 对象*/
7.          JdbcTemplate jdbcTemplate =
8.                  (JdbcTemplate)applicationContext.getBean("jdbcTemplate");
9.
10.         /*进行增删改操作*/
11.         String sql1 = "update stu set score = 60 where id = 1";
12.         String sql2 = "insert into stu(name,course,score) " +
13.                 "values('赵六','英语',89)";
14.         String sql3 = "delete from stu where id = 3";
15.
```

```
16.         int dex1 = jdbcTemplate.update(sql1);
17.         int dex2 = jdbcTemplate.update(sql2);
18.         int dex3 = jdbcTemplate.update(sql3);
19.         /*dex 表示此次更新影响的行数，其值为 0 时表示此次更新无效*/
20.         System.out.println(dex1+"-"+dex2+"-"+dex3);
21.     }
22. }
```

在例 5-7 中，JdbcTemplate 的 update()方法负责进行数据的增删改操作，将所需的 SQL 语句作为参数放入其中即可，其返回值为此次更新影响的行数。若返回值为 0，则此次更新无效。例 5-7 代码的运行结果如下所示。

```
1-1-1
```

5.2.4 DDL 操作

利用 execute()方法可完成 DDL 操作。下面使用一个示例进行讲解，如例 5-8 所示。

【例 5-8】 Test.java

```
1.  public class Test {
2.      public static void main(String[] args) {
3.          ApplicationContext applicationContext =
4.              new ClassPathXmlApplicationContext("applicationContext.xml");
5.
6.          /*获取 jdbcTemplate 对象*/
7.          JdbcTemplate jdbcTemplate =
8.                  (JdbcTemplate)applicationContext.getBean("jdbcTemplate");
9.
10.         /*创建老师表 teacher*/
11.         String sql = "CREATE TABLE 'teacher'(\n" +
12.                 "'id' int(0) NOT NULL AUTO_INCREMENT,\n" +
13.                 "'name' varchar(255),\n" +
14.                 " 'course' varchar(255),\n" +
15.                 " PRIMARY KEY ('id') USING BTREE\n" +
16.                 " )";
17.         /*此方法没有返回值*/
18.         jdbcTemplate.execute(sql);
19.     }
20. }
```

在例 5-8 中，使用 execute()方法可执行创建表语句、更改表结构语句等 SQL 语句，此方法没有返回值，需要通过连接工具查看结果。本示例选用的连接工具是 Navicat，创建结果如图 5.4 所示。

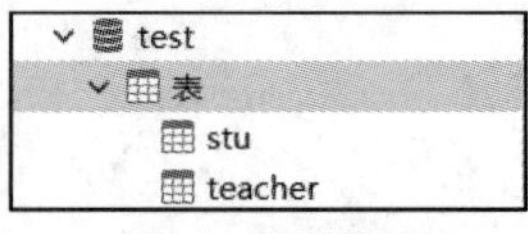

图 5.4 创建结果

5.3 JdbcTemplate 在日常开发中的使用

本节展示 JdbcTemplate 在日常开发中的使用。在此使用例 5-1 中的 applicationContext.xml 文件作为 Spring 的配置文件，编写业务代码，具体代码如例 5-9 所示。

【例 5-9】 Test.java

```
1.  @Component
```

```
2.  public class StudentService{
3.
4.      @Autowired
5.      JdbcTemplate jdbcTemplate;
6.
7.      public void getStudent(){
8.          List<Student> student =
9.                  jdbcTemplate.query(
10.                         "select * from stu where id=1",
11.                         new BeanPropertyRowMapper<>(Student.class));
12.     }
13. }
```

在例 5-9 中，使用@Autowired 注解直接取出容器中的 JdbcTemplate 对象，随后使用 JdbcTemplate 对象进行查询操作。

5.4 本章小结

本章介绍了 Spring 与数据库交互的相关知识，其中包括数据库的配置和使用 JdbcTemplate 操作数据库。通过对本章的学习，读者可以初步掌握操作数据库的方法，为下一章 Spring 事务的学习做铺垫。

5.5 习题

1．填空题

（1）在 Spring JDBC 中可以通过工具类________直接操作数据库，使用此工具类需要配置数据库连接源________来连接数据库。

（2）JdbcTemplate 类中更改和查询的方法分别为________和________。

2．选择题

（1）关于 Spring JDBC，下列描述错误的是（　　）。

A. Spring JDBC 对传统 JDBC 进行了改善和增强

B. Spring JDBC 和传统 JDBC 没有关联

C. Spring JDBC 的 core 包负责提供核心功能

D. Spring JDBC 的 support 包负责提供支持类

（2）在下列选项中，不属于数据源连接配置的参数是（　　）。

A. driverClassName　B. user　C. username　D. password

（3）在 JdbcTemplate 类提供的方法中，用于查询单条记录的是（　　）。

A. batchUpdate()　B. queryForList()　C. queryForObject()　D. update()

3．编程题

创建 Dog 类，在其中添加 name 和 age 属性，并创建对应数据库，利用 JdbcTemplate 完成对 Dog 类中数据的增删查改操作。

第6章 Spring 事务

本章学习目标

- 理解事务的概念。
- 理解事务的管理方式和事务失效问题。
- 掌握声明式事务管理的实现方法。
- 掌握事务的传播方式。

Spring 事务

通过对第 5 章的学习，读者已经了解了如何使用 JdbcTemplate 操作数据库，本章介绍 Spring 事务。在日常开发中，事务的操作非常繁杂，Spring 提供了专门负责事务处理的 API，减少了开发人员的工作量。Spring 事务是 Spring 进阶内容中较复杂且重要的核心知识点。下面对 Spring 的事务管理进行详解讲解。读者需要深入理解事务的概念，熟练掌握基于注解实现声明式事务管理的方法，初步掌握事务的传播方式。

6.1 事务概述

事务是指业务上不可分割的一段代码，此段业务代码只有全部执行成功和全部执行失败两种执行结果。Spring 为开发人员提供了事务管理的核心接口。本节将从示例着手，讲解事务的概念与事务的管理方式，读者需要初步理解事务与声明式事务的概念。

6.1.1 事务管理

在日常开发中，通常使用 Spring 中的@Transaction 注解处理事务，下面以一段代码为例讲解 Spring 处理事务的常用方式，如例 6-1 所示。

【例 6-1】 ModifyAccount.java

```
1.  //将 ModifyAccount 交由容器管理
2.  @Component
3.  public class ModifyAccount {
4.      /*从容器中取出 jdbcTemplate 对象*/
5.      @Autowired
6.      JdbcTemplate jdbcTemplate;
7.
8.      @Transactional
9.      public void transAccounts(){
10.         /*将张三账户中的金额增加 10*/
11.         jdbcTemplate.update(
12.                 "update account set money = 20 where name = '张三'");
```

```
13.        /*模拟错误*/
14.        int a = 0/0;
15.        /*将李四账户中的金额减少 10*/
16.        jdbcTemplate.update(
17.               "update account set money = 30 where name = '李四'");
18.    }
19. }
```

在例 6-1 中，在 transAccounts()转账方法上方添加@Transactional 注解，将该方法的内容作为一个事务处理，其中表 account 的结构与数据如图 6.1 所示。

id	name	money
1	张三	10
2	李四	40

图 6.1 表 account 的结构与数据

除此之外，还需要配置 applicationContext.xml 文件，在其中启用事务注解，如例 6-2 所示。

【例 6-2】 applicationContext.xml

```
1.  <?xml version="1.0" encoding="UTF-8"?>
2.  <beans xmlns="http://www.springframework.org/schema/beans"
3.       xmlns:xsi="http://www.w3.org/2001/XMLSchema-instance"
4.       xmlns:context="http://www.springframework.org/schema/context"
5.       xmlns:aop="http://www.springframework.org/schema/aop"
6.       xmlns:tx="http://www.springframework.org/schema/tx"
7.       xsi:schemaLocation="http://www.springframework.org/schema/beans
8.        http://www.springframework.org/schema/beans/spring-beans.xsd
9.        http://www.springframework.org/schema/context
10.       http://www.springframework.org/schema/context/spring-context.xsd
11.       http://www.springframework.org/schema/aop
12.       http://www.springframework.org/schema/aop/spring-aop-4.3.xsd
13.       http://www.springframework.org/schema/tx
14.       http://www.springframework.org/schema/tx/spring-tx.xsd">
15.    <!--扫描 com 包下的所有注解-->
16.    <context:component-scan base-package="com"/>
17.
18.    <!--配置连接-->
19.    <bean id="myDataSource"
20.     class="org.springframework.jdbc.datasource.DriverManagerDataSource">
21.       <!--连接驱动-->
22.       <property name="driverClassName" value="com.mysql.cj.jdbc.Driver"/>
23.       <!--连接地址-->
24.       <property
25.        name="url"
26.        value="jdbc:mysql://127.0.0.1/test?characterEncoding=utf8&amp;
27.               useSSL=false&amp;
28.               serverTimezone=UTC&amp;
29.               allowPublicKeyRetrieval=true"/>
30.       <!--用户名-->
31.       <property name="username" value="root"/>
32.       <!--密码-->
33.       <property name="password" value="******"/>
34.    </bean>
35.
36.    <!--配置模板-->
37.    <bean id="jdbcTemplate"
```

```
38.         class="org.springframework.jdbc.core.JdbcTemplate">
39.       <!--配置数据源-->
40.       <property name="dataSource" ref="myDataSource"/>
41.    </bean>
42.    <!--配置事务管理器-->
43.    <bean id="transactionManager"
44. class="org.springframework.jdbc.datasource.DataSourceTransactionManager">
45.       <property name="dataSource" ref="myDataSource"/>
46.    </bean>
47.    <!--启用注解-->
48.    <tx:annotation-driven transaction-manager="transactionManager"/>
49. </beans>
```

在例 6-2 中，首先在代码的第 6、13、14 行添加 tx 命名空间，然后配置数据库连接 myDataSource 与数据库连接模板 jdbcTemplate，最后配置事务管理器 transactionManager，此事务管理器负责管理数据库的增删改查事务，需要传入 data source 作为参数。配置完成后，使用 tx:annotation-driven 开启事务注解扫描。

在 Test 类中，获取 modifyAccount 对象，调用 transAccounts()方法，Test 类代码如例 6-3 所示。

【例 6-3】 Test.java

```
1.  public class Test {
2.     public static void main(String[] args) {
3.        ApplicationContext applicationContext =
4.              new ClassPathXmlApplicationContext("applicationContext.xml");
5.
6.        /*获取 modifyAccount 对象*/
7.        ModifyAccount modifyAccount =
8.               (ModifyAccount)applicationContext.getBean("modifyAccount");
9.        /*调用 transAccounts()方法*/
10.       modifyAccount.transAccounts();
11.    }
12. }
```

执行例 6-3 中的代码，调用 transAccounts()方法，Spring 事务开启，transAccounts()方法中张三的金额新增代码执行完毕，之后遇到错误，李四的金额减少代码没有执行，此时 Spring 发现此事务中存在错误，回滚此事务。最后，张三的金额并没有增加，保证了系统的安全。

6.1.2 事务的管理方式

事务的管理方式有很多种，在此简单介绍其中的两种，分别为编程式事务管理和声明式事务管理，下面详细介绍这两种事务管理方式的区别。

1．编程式事务管理

编程式事务管理是指通过编写代码的方式实现事务管理，是最基础的事务管理方式。由于其操作较烦琐，在开发中不经常使用，在此举例简单说明，读者了解即可，具体代码如例 6-4 所示。

【例 6-4】 ModifyAccount.java

```
1.  //将 ModifyAccount 交由容器管理
2.  @Component
3.  public class ModifyAccount {
```

```
4.      /*从容器中取出 jdbcTemplate 对象*/
5.      @Autowired
6.      JdbcTemplate jdbcTemplate;
7.
8.      /*从容器中取出 transactionTemplate 对象*/
9.      @Autowired
10.     TransactionTemplate transactionTemplate;
11.
12.     public void transAccounts(){
13.        transactionTemplate.execute(new TransactionCallbackWithoutResult() {
14.            /*在事务环境中执行的代码*/
15.            @Override
16.            protected void doInTransactionWithoutResult(
17.                                        TransactionStatus status) {
18.                /*将张三账户中的金额增加 10*/
19.                jdbcTemplate.update(
20.                        "update account set money = 20 where name = '张三'");
21.                /*模拟错误*/
22.                int a = 0/0;
23.                /*将李四账户中的金额减少 10*/
24.                jdbcTemplate.update(
25.                        "update account set money = 30 where name = '李四'");
26.            }
27.        });
28.
29.     }
30. }
```

jdbcTemplate 是负责数据库交互的工具，而 transactionTemplate 是实现事务管理的工具。TransactionTemplate 类需要在 Spring 配置文件中注入，代码如例 6-5 所示。

【例 6-5】 applicationContext.java

```
1.  <bean  id="transactionTemplate"
2.     class="org.springframework.transaction.support.TransactionTemplate">
3.         <property name="transactionManager" ref="transactionManager"/>
4.  </bean>
```

例 6-5 中的 TransactionTemplate 需要事务管理器 transactionManager 作为参数，配置完成后即可注入。之后调用例 6-4 中 TransactionTemplate 的 execute()方法，此方法需要传入参数类型为 TransactionCallbackWithoutResult 的类。此类是 TransactionCallback 的实现类，提供了 doInTransactionWithoutResult()方法，该方法用于调用在事务环境中执行的代码。将被事务管理的代码放入 doInTransactionWithoutResult()中，即可实现事务的管理。

2．声明式事务管理

声明式事务管理是通过 Spring AOP 技术实现的，此管理方式将与事务相关的代码（如开启事务和回滚事务等代码）作为切面，织入目标方法上，即在目标方法前后加上事务管理的语句。当目标方法中的代码发生异常时，Spring AOP 在“方法结束时”连接点处回滚此事务，以此达到事务管理的目的。

此方式简单便捷，代码的侵入性较低，是开发时常用的事务管理方法。通常，声明式事务管理有两种实现方式：基于注解和基于 XML。下面讲解声明式事务管理的实现方法。

6.2 声明式事务管理的实现方法

本节将介绍常用声明式事务管理的实现方法，其中，基于注解实现声明式事务管理需要读者熟练掌握，基于 XML 实现声明式事务管理需要读者了解。

6.2.1 基于注解实现声明式事务管理

6.1.1 小节中的示例基于注解完成了声明式事务。要实现基于注解的声明式事务管理，只需以下两个步骤。

（1）在 Spring 配置文件中启用事务注解，代码如下所示。

```
<tx:annotation-driven transaction-manager="transactionManager"/>
```

（2）在需要进行事务处理的类或方法上方加上@Transaction 注解。在此需要注意的是，如果此注解加到某类上方，则表明此类的所有方法都将被事务处理；如果此注解加到某方法上方，则表示此方法将使用事务来处理；如果在一个方法与类上方都加上@Transaction 注解，则被此注解标注的方法会使用全新的事务处理，其他未被此注解标注的方法使用被此注解标注的类的事务处理。

@Transaction 注解的属性如表 6.1 所示。

表 6.1 @Transaction 注解的属性

属性名称	说明
value	指定一个事务管理器，默认为 transactionManager
transactionManager	当配置多个事务管理器时，可以用于选择使用哪个事务管理器
isolation	指定事务的隔离级别
noRollbackFor	指定事务遇到特定异常时不回滚，并指定异常类 class 数组
noRollbackForClassName	指定事务遇到特定异常时不回滚，可以指定多个异常类名
propagation	指定事务的传播方式
readOnly	指定事务是否为只读，默认为 false
rollbackFor	指定事务遇到特定异常时回滚，并指定异常类 class 数组
rollbackForClassName	指定事务遇到特定异常时回滚，可以指定多个异常类名
timeout	指定事务的超时时长

propagation 属性将在 6.3 节做详细介绍，在此讲解 readOnly 和 timeout 属性的配置，如例 6-6 所示。

【例 6-6】 ModifyAccount.java

```
1.  //将 ModifyAccount 交由容器管理
2.  @Component
3.  public class ModifyAccount {
4.      /*从容器中取出 jdbcTemplate 对象*/
5.      @Autowired
6.      JdbcTemplate jdbcTemplate;
7.
8.      @Transactional(readOnly = true)
9.      public void transAccounts1(){
10.         /*模拟修改，将张三账户中的金额增加 10*/
11.         jdbcTemplate.update(
```

```
12.             "update account set money = 30 where name = '张三'");
13.     }
14.
15.     @Transactional(timeout = 1)
16.     public void transAccounts2() throws InterruptedException {
17.         /*模拟超时*/
18.         Thread.sleep(3000);
19.         /*将张三账户中的金额增加 10*/
20.         jdbcTemplate.update(
21.                 "update account set money = 10 where name = '张三'");
22.     }
23. }
```

在例 6-6 中，第 8 行代码将 readOnly 属性的值设置为 true，表示此事务为只读。而在此事务方法中，执行了数据库的数据更改操作，因此回滚此事务，并且抛出异常。编写测试类并执行 transAccounts1()方法，结果如图 6.2 所示。

```
Caused by: java.sql.SQLException Create breakpoint : Connection is read-only. Queries leading to data modification are not allowed.
    at com.mysql.cj.jdbc.exceptions.SQLError.createSQLException(SQLError.java:129)
    at com.mysql.cj.jdbc.exceptions.SQLError.createSQLException(SQLError.java:97)
    at com.mysql.cj.jdbc.exceptions.SQLError.createSQLException(SQLError.java:89)
```

图 6.2 transAccounts1()方法的执行结果

在例 6-6 中，第 15 行代码将 timeout 属性的值设置为 1，表示此事务的超时时间为 1 秒。而在此事务方法中，执行时间超过了 1 秒，因此回滚此事务，并且抛出异常。编写测试类并执行 transAccounts2()方法，结果如图 6.3 所示。

```
Exception in thread "main" org.springframework.transaction.TransactionTimedOutException Create breakpoint : Transaction timed out: deadline was Wed Nov 03 16:38:36 CST 2021
    at org.springframework.transaction.support.ResourceHolderSupport.checkTransactionTimeout(ResourceHolderSupport.java:155)
    at org.springframework.transaction.support.ResourceHolderSupport.getTimeToLiveInMillis(ResourceHolderSupport.java:144)
    at org.springframework.transaction.support.ResourceHolderSupport.getTimeToLiveInSeconds(ResourceHolderSupport.java:128)
    at org.springframework.jdbc.datasource.DataSourceUtils.applyTimeout(DataSourceUtils.java:288)
```

图 6.3 transAccounts2()方法的执行结果

6.2.2 基于 XML 实现声明式事务管理

下面讲解基于 XML 方式完成声明式事务管理的方法。首先需要了解一些声明式事务的属性，如表 6.2 所示。

表 6.2 声明式事务的属性

属性名称	说明
name	指定一个事务管理器，默认为 transactionManager
isolation	指定事务的隔离级别
no-rollback-for	指定事务遇到特定异常时不回滚，并指定异常类 class 数组
propagation	指定事务的传播方式
readOnly	指定事务是否为只读，默认为 false
rollback-for	指定事务遇到特定异常时回滚，并指定异常类 class 数组
timeout	指定事务的超时时长

表 6.2 中的属性与@Tramsaction 注解的部分属性相同。下面使用示例讲解基于 XML 完成声明式事务管理的方法。创建目录结构，目录结构如图 6.4 所示。

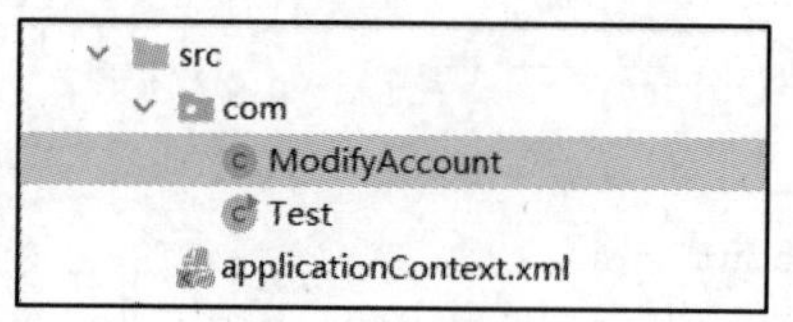

图 6.4　目录结构

在 com 包下创建 ModifyAccount 目标类和 Test 类，编写 Spring 配置文件。下面通过示例了解配置文件中与事务相关的元素，配置文件如例 6-7 所示。

【例 6-7】 applicationContext.xml

```
1.  <?xml version="1.0" encoding="UTF-8"?>
2.  <beans xmlns="http://www.springframework.org/schema/beans"
3.         xmlns:xsi="http://www.w3.org/2001/XMLSchema-instance"
4.         xmlns:context="http://www.springframework.org/schema/context"
5.         xmlns:aop="http://www.springframework.org/schema/aop"
6.         xmlns:tx="http://www.springframework.org/schema/tx"
7.         xsi:schemaLocation="http://www.springframework.org/schema/beans
8.           http://www.springframework.org/schema/beans/spring-beans.xsd
9.           http://www.springframework.org/schema/context
10.          http://www.springframework.org/schema/context/spring-context.xsd
11.          http://www.springframework.org/schema/aop
12.          http://www.springframework.org/schema/aop/spring-aop-4.3.xsd
13.          http://www.springframework.org/schema/tx
14.          http://www.springframework.org/schema/tx/spring-tx.xsd">
15.     <!--扫描 com 包下的所有注解-->
16.     <context:component-scan base-package="com"/>
17.
18.     <!--配置连接-->
19.     <bean id="myDataSource"
20.       class="org.springframework.jdbc.datasource.DriverManagerDataSource">
21.         <!--连接驱动-->
22.         <property name="driverClassName" value="com.mysql.cj.jdbc.Driver"/>
23.         <!--连接地址-->
24.         <property
25.          name="url"
26.          value="jdbc:mysql://127.0.0.1/test?characterEncoding=utf8&amp;
27.                useSSL=false&amp;
28.                serverTimezone=UTC&amp;
29.                allowPublicKeyRetrieval=true"/>
30.         <!--用户名-->
31.         <property name="username" value="root"/>
32.         <!--密码-->
33.         <property name="password" value="2865gfv79349"/>
34.     </bean>
35.
36.     <!--配置模板-->
37.     <bean id="jdbcTemplate"
38.           class="org.springframework.jdbc.core.JdbcTemplate">
39.         <!--配置数据源-->
40.         <property name="dataSource" ref="myDataSource"/>
41.     </bean>
42.
```

```
43.     <!--配置事务管理器-->
44.     <bean id="transactionManager"
45. class="org.springframework.jdbc.datasource.DataSourceTransactionManager">
46.         <property name="dataSource" ref="myDataSource"/>
47.     </bean>
48.
49.     <!--编写通知-->
50.     <tx:advice
51.             id="txAdvice"
52.             transaction-manager="transactionManager">
53.         <tx:attributes>
54.             <!--配置属性-->
55.             <tx:method name="transAccounts1"
56.                     propagation="REQUIRED"
57.                     isolation="DEFAULT"
58.                     timeout="-1"
59.                     readOnly="false"/>
60.         </tx:attributes>
61.     </tx:advice>
62.     <!--将通知映射到切点上-->
63.     <aop:config>
64.         <!--通过 Aop 使用 advice-ref 对切点进行增强操作-->
65.         <aop:pointcut id="transM"
66.           expression="execution(* com.ModifyAccount.transAccounts1(..))"/>
67.         <aop:advisor advice-ref="txAdvice" pointcut-ref="transM"/>
68.     </aop:config>
69. </beans>
```

在例 6-7 中，<tx:advice>元素用于声明一个通知，并在其中配置<tx:attributes>和<tx:method>元素。<tx:method>元素为核心元素，代表一个方法，在此元素中使用 name 属性指定方法名，并使用表 6.2 中的相关属性配置此方法的事务。

配置完声明式事务后，通过 AOP 配置切面到切点的映射，切点为 ModifyAccount 类中的 transAccounts1()方法，目标类 ModifyAccount 如例 6-8 所示。

【例 6-8】 ModifyAccount.java

```
1.  //将 ModifyAccount 交由容器管理
2.  @Component
3.  public class ModifyAccount {
4.      /*从容器中取出 jdbcTemplate 对象*/
5.      @Autowired
6.      JdbcTemplate jdbcTemplate;
7.
8.      public void transAccounts1(){
9.          /*模拟修改，将李四账户中的金额减少 10*/
10.         jdbcTemplate.update(
11.                 "update account set money = 30 where name = '李四'");
12.         /*模拟错误*/
13.         int a = 1/0;
14.         /*模拟修改，将张三账户中的金额增加 10*/
15.         jdbcTemplate.update(
16.                 "update account set money = 20 where name = '张三'");
```

```
17.     }
18. }
```

在目标类 ModifyAccount 中，模拟李四向张三转账，转账后李四账户的金额为 40 元，张三账户的金额为 10 元。先减少李四账户的金额，再增加张三账户的金额，在减少李四账户的金额后报错。然后通过 Test 类查看最后结果，Test 类如例 6-9 所示。

【例 6-9】 Test.java

```
1.  public class Test {
2.
3.      public static void main(String[] args) throws InterruptedException {
4.          ApplicationContext applicationContext =
5.                  new ClassPathXmlApplicationContext("applicationContext.xml");
6.
7.          /*获取 modifyAccount 对象*/
8.          ModifyAccount modifyAccount =
9.                  (ModifyAccount)applicationContext.getBean("modifyAccount");
10.         /*调用 transAccounts1()方法*/
11.         modifyAccount.transAccounts1();
12.     }
13. }
```

在例 6-9 中，通过 getBean()方法获取容器中的 modifyAccount 对象，然后调用 transAccounts1()方法，执行 Test 类后的数据库如图 6.5 所示。

从图 6.5 中可以看到，李四账户的金额并没有改变，事务成功回滚。

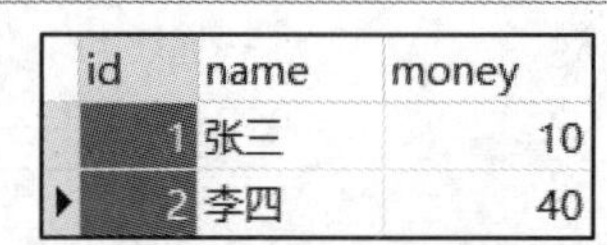

id	name	money
1	张三	10
2	李四	40

图 6.5　执行 Test 类后的数据库

6.3 事务的传播方式

当一个事务方法（用@Transactional 注解标记的方法）调用另一个事务方法时，事务之间会出现冲突，因为方法无法选择使用哪个事务。为解决此类问题，Spring 提供了 7 种事务传播方式。本节将详细讲解 7 种传播方式，读者需要熟练掌握常用的事务传播方式。

下面将用示例讲解各种事务的传播方式，主要代码如例 6-10 所示。

【例 6-10】 Test.java

```
1.  public void main() {
2.      insertA(a); //将 a 插入 A 数据库
3.      test();
4.  }
5.  //其他类中的 test()方法
6.  public void test(){
7.      insertB(b); //将 b 插入 B 数据库
8.      throw Exception;
9.      insertB(c); //将 c 插入 B 数据库
10. }
```

例 6-10 所示的是 main()方法与 test()方法，这两个方法在不同的类中。通过 main()方法将 a 插入 A 数据库，随后跨类调用 test()方法，在通过 test()方法将 b 插入 B 数据库后，抛出异常，然后将 c 插入 B 数据库。现在为这两个方法加上事务，通过使用不同的传播方式来分析事务的执行情况。

1. PROPAGATION_REQUIRED（默认事务传播方式）

PROPAGATION 译为传播方式，REQUIRED 译为需要，在后面的讲解中，为了方便，把 PROPAGATION 省略，将该传播方式称为 REQUIRED 传播方式。此传播方式的作用为：如果当前没有事务，则新建一个事务；如果当前存在事务，则加入这个事务。下面阅读例 6-11 的代码，思考其运行结果。

【例 6-11】 Test.java

```
1.  @Transactional(propagation = Propagation.REQUIRED)
2.  public void main() {
3.      insertA(a); //将a插入A数据库
4.      try{
5.        test();
6.      }catch(){}
7.  }
8.  @Transactional(propagation = Propagation.REQUIRED)
9.  public void test(){
10.     insertB(b); //将b插入B数据库
11.     throw Exception; //抛出异常
12.     insertB(c); //将c插入B数据库
13. }
```

例 6-11 代码的运行结果为：没有一个插入操作被执行。下面分析上述代码的运行流程。

（1）main()方法被执行，main()方法的事务传播方式为 REQUIRED，此时没有事务，则新建一个事务。

（2）执行插入操作，将 a 插入 A 数据库，然后调用 test()方法。

（3）执行 test()方法，test()方法的事务传播方式为 REQUIRED，此时存在 main()方法的事务，则使用 main()方法的事务。

（4）执行插入操作，将 b 插入 B 数据库，随后抛出异常，回滚事务，此时使用的事务为 main()方法的事务，因此回滚 b 与 a 的插入操作，结束程序。

需要注意的是，虽然在 main()方法中使用 try catch 处理了异常，使得 main()方法中不存在异常，但是由于 test()方法使用了 main()方法的事务，在事务里抛出了异常，因此会直接回滚此事务中的全部内容，在后续的 NESTED 传播方式的讲解中会与此例做对比。

2. PROPAGATION_REQUIRES_NEW

REQUIRES_NEW 为需要新事务，此传播方式的作用为：如果当前没有事务，则新建一个事务；如果当前存在事务，仍然新建一个事务。下面阅读例 6-12 中的代码，思考其运行结果。

【例 6-12】 Test.java

```
1.  @Transactional(propagation = Propagation.REQUIRED)
2.  public void main() {
3.      insertA(a); //将a插入A数据库
4.      test();
5.      throw Exception; //抛出异常
6.  }
7.  @Transactional(propagation = Propagation.REQUIRED_NEW)
8.  public void test(){
9.      insertB(b); //将b插入B数据库
10.     insertB(c); //将c插入B数据库
11. }
```

例 6-12 中代码的运行结果为：b 和 c 插入了 B 数据库。下面分析上述代码的运行流程。

（1）main()方法被执行，main()方法中事务的传播方式为 REQUIRED，此时没有事务，则新建一个事务。

（2）执行插入操作，将 a 插入 A 数据库，然后调用 test()方法。

（3）执行 test()方法，test()方法的事务传播方式为 REQUIRED_NEW，此时新建一个事务。

（4）执行插入操作，将 b、c 插入 B 数据库，test()方法执行结束，test()方法的事务提交。此时 test()方法的事务中存在 b 与 c 的插入操作，提交这两个操作，返回。

（5）随后，main()方法抛出异常，因为此时使用的事务为 main()方法的事务，所以回滚 a 的插入操作，结束程序。

3．PROPAGATION_SUPPORTS

SUPPORTS 为支持事务，此传播方式的作用为：如果当前没有事务，就不用事务；如果当前存在事务，则使用此事务。下面阅读例 6-13 中的代码，思考其运行结果。

【例 6-13】 Test.java

```
1.  public void main() {
2.      insertA(a); //将a插入A数据库
3.      test();
4.  }
5.  @Transactional(propagation = Propagation.SUPPORTS)
6.  public void test(){
7.      insertB(b); //将b插入B数据库
8.      throw Exception; //抛出异常
9.      insertB(c); //将c插入B数据库
10. }
```

例 6-13 中代码的运行结果为：a 和 b 分别插入了 A 和 B 数据库。下面分析上述代码的运行流程。

（1）main()方法被执行，main()方法没有事务，执行插入操作，将 a 插入 A 数据库，然后调用 test()方法。

（2）执行 test()方法，test()方法的事务传播方式为 SUPPORTS，因此不使用事务。

（3）执行插入操作，将 b 插入 B 数据库，随后抛出异常，程序中断执行，结束程序。

4．PROPAGATION_NOT_SUPPORTED

NOT_SUPPORTED 为不支持事务，此传播方式的作用为：始终以非事务方式执行，如果当前存在事务，则挂起当前事务。下面阅读例 6-14 中的代码，思考其运行结果。

【例 6-14】 Test.java

```
1.  @Transactional(propagation = Propagation.REQUIRED)
2.  public void main() {
3.      insertA(a); //将a插入A数据库
4.      test();
5.  }
6.  @Transactional(propagation = Propagation.NOT_SUPPORTED)
7.  public void test(){
8.      insertB(b); //将b插入B数据库
9.      throw Exception; //抛出异常
10.     insertB(c); //将c插入B数据库
11. }
```

例 6-14 中代码的运行结果为：只有 b 插入了 B 数据库。下面分析上述代码的运行流程。

（1）main()方法被执行，main()方法的事务传播方式为 REQUIRED，此时没有事务，则新建一个事务。

（2）执行插入操作，将 a 插入 A 数据库，然后调用 test()方法。

（3）执行 test()方法，test()方法的事务传播方式为 NOT_SUPPORTED，此时挂起 main()方法的事务，以非事务方式运行。

（4）执行插入操作，将 b 插入 B 数据库，随后抛出异常，中止程序。此时 main()方法的事务接收到抛出的异常，事务内只存在 a 的插入操作，不存在 b 的插入操作，因此回滚 a 的插入操作，结束程序。

5．PROPAGATION_MANDATORY

MANDATORY 为强制使用事务运行，此传播方式的作用为：如果当前存在事务，则使用当前事务；如果当前不存在事务，则直接抛出异常。下面阅读例 6-15 中的代码，思考其运行结果。

【例 6-15】 Test.java

```
1.  public void main() {
2.      insertA(a); //将 a 插入 A 数据库
3.      test();
4.  }
5.  @Transactional(propagation = Propagation.MANDATORY)
6.  public void test(){
7.      insertB(b); //将 b 插入 B 数据库
8.      throw Exception; //抛出异常
9.      insertB(c); //将 c 插入 B 数据库
10. }
```

例 6-15 中代码的运行结果为：只有 a 插入了 A 数据库。下面分析上述代码的运行流程。

（1）main()方法被执行，main()方法没有事务，执行插入操作，将 a 插入 A 数据库，然后调用 test()方法。

（2）执行 test()方法，test()方法的事务传播方式为 MANDATORY，此时没有事务，抛出异常，终止程序。

6．PROPAGATION_NEVER

NEVER 为绝对不使用事务，此传播方式的作用为：不使用事务，如果当前存在事务，则抛出异常。下面阅读例 6-16 中的代码，思考其运行结果。

【例 6-16】 Test.java

```
1.  @Transactional(propagation = Propagation.REQUIRED)
2.  public void main() {
3.      insertA(a); //将 a 插入 A 数据库
4.      test();
5.  }
6.  @Transactional(propagation = Propagation.NEVER)
7.  public void test(){
8.      insertB(b); //将 b 插入 B 数据库
9.      throw Exception; //抛出异常
10.     insertB(c); //将 c 插入 B 数据库
11. }
```

例 6-16 中代码的运行结果为：只有 a 插入了 A 数据库。下面分析上述代码的运行流程。

（1）main()方法被执行，main()方法没有事务，执行插入操作，将 a 插入 A 数据库，然后调用 test()方法。

（2）执行 test()方法，test()方法的事务传播方式为 NEVER，此时存在 main()方法的事务，抛出异常，终止程序。

7．PROPAGATION_NESTED

NESTED 为嵌套事务，此传播方式的作用为：如果当前存在事务，则在嵌套事务中执行，否则新建一个事务。在嵌套事务中，某事务虽然在原事务中执行，但是在此嵌套事务中发生的异常可以被捕获，不影响原事务的执行。阅读例 6-17 中的代码，并将其与例 6-11 做对比，思考其运行结果。

【例 6-17】 Test.java

```
1.  @Transactional(propagation = Propagation.REQUIRED)
2.  public void main() {
3.      insertA(a); //将 a 插入 A 数据库
4.      try{
5.         test();
6.      }catch(){}
7.  }
8.  @Transactional(propagation = Propagation.NESTED)
9.  public void test(){
10.     insertB(b); //将 b 插入 B 数据库
11.     throw Exception; //抛出异常
12.     insertB(c); //将 c 插入 B 数据库
13. }
```

例 6-17 的运行结果为：只有 a 插入了 A 数据库。下面分析上述代码的运行流程。

（1）main()方法被执行，main()方法的事务传播方式为 REQUIRED，此时没有事务，则新建一个事务。

（2）执行插入操作，将 a 插入 A 数据库，然后调用 test()方法。

（3）执行 test()方法，test()方法的事务传播方式为 NESTED，此时在嵌套事务中运行代码。

（4）执行插入操作，将 b 插入 B 数据库，随后抛出异常，回滚此嵌套事务，向外抛出异常。

（5）异常被捕获，main()方法继续执行，main()方法执行结束后，事务结束，提交操作，此时 main()方法的事务中存在 a 的插入操作，执行此操作，结束程序。

为了区分 NESTED 与 REQUIRED_NEW 这两种传播方式，阅读例 6-18 中的代码，分析其运行结果。

【例 6-18】 Test.java

```
1.  @Transactional(propagation = Propagation.REQUIRED)
2.  public void main() {
3.      insertA(a); //将 a 插入 A 数据库
4.      test();
5.      throw Exception; //抛出异常
6.  }
7.  @Transactional(propagation = Propagation.NESTED)
8.  public void test(){
9.      insertB(b); //将 b 插入 B 数据库
10.     insertB(c); //将 c 插入 B 数据库
11. }
```

例 6-18 中代码的运行结果为：没有操作被执行。下面分析上述代码的运行流程。

（1）main()方法被执行，main()方法的事务传播方式为 REQUIRED，此时没有事务，则新建一个事务。

（2）执行插入操作，将 a 插入 A 数据库，然后调用 test()方法。

（3）执行 test()方法，test()方法的事务传播方式为 NESTED，此时新建一个嵌套事务。

（4）执行插入操作，将 b、c 插入 B 数据库，test()方法执行结束，test()方法的事务嵌套在 main()方法的事务中，必须随着 main()方法的事务一同提交，因此直接返回。

（5）随后，main()方法抛出异常，当前使用的事务为 main()方法的事务，因此回滚 a 的插入操作与嵌套事务中 b、c 的插入操作，结束程序。

简单总结 NESTED 传播方式的要点：嵌套事务中的异常可以被原事务捕捉，从而不影响原事务的运行，而在原事务发生异常后，嵌套事务会随着原事务一起回滚。

6.4 事务失效问题

在日常开发中经常会遇到事务失效问题，本节将介绍事务失效的原因，读者可初步了解。

1．同一个类中的事务方法之间的调用

在同一个类中，非事务方法调用事务方法会导致事务方法失效，如例 6-19 所示。

【例 6-19】 ModifyAccount.java

```
1.  //将 ModifyAccount 交由容器管理
2.  @Component
3.  public class ModifyAccount {
4.      /*从容器中取出 jdbcTemplate 对象*/
5.      @Autowired
6.      JdbcTemplate jdbcTemplate;
7.
8.      public void test(){
9.          transAccounts1();
10.     }
11.
12.     @Transactional(propagation = Propagation.REQUIRED)
13.     public void transAccounts1(){
14.         /*模拟修改，将李四账户中的金额减少 10*/
15.         jdbcTemplate.update(
16.                 "update account set money = 30 where name = '李四'");
17.         /*模拟错误*/
18.         int a = 1/0;
19.         /*模拟修改，将张三账户中的金额增加 10*/
20.         jdbcTemplate.update(
21.                 "update account set money = 20 where name = '张三'");
22.     }
23. }
```

调用 test()方法，在 test()方法内部调用同一个类中的事务方法 transAccounts1()，程序的运行结果为：李四账户中的金额减少，而张三账户中的金额没有增加，此事务失效。

产生这个问题的关键在于，事务是基于 AOP 实现的，用容器获取的类其实是 IoC 容器创建的代理类，调用此类的 test()方法，实际调用的是被代理类中的 test()方法，而在被代理类的 test()方法中，调用的 transAccounts1()方法实际上隐藏了 this，即调用的是 this.transAccounts1()，之后被代理类中的 transAccounts1()方法被调用，此方法没有被 Spring 事务管理，因此事务失效。

2．异常被捕获

如果在程序的运行过程中异常被捕获，则该事务将无法检测到错误，不会回滚数据，如例 6-20 所示。

【例 6-20】 ModifyAccount.java

```
1.  public class ModifyAccount {
2.      /*从容器中取出 jdbcTemplate 对象*/
3.      @Autowired
4.      JdbcTemplate jdbcTemplate;
5.
6.      @Transactional(propagation = Propagation.REQUIRED)
7.      public void transAccounts1(){
8.          /*模拟修改，将李四账户中的金额减少 10*/
9.          jdbcTemplate.update(
10.                 "update account set money = 30 where name = '李四'");
11.         try {
12.             /*模拟错误*/
13.             int a = 1/0;
14.         } catch (Exception e) {
15.             System.out.println("异常被捕获");
16.         }
17.         /*模拟修改，将张三账户中的金额增加 10*/
18.         jdbcTemplate.update(
19.                 "update account set money = 20 where name = '张三'");
20.     }
21. }
```

此错误容易理解，但是并不容易避免，在开发中预料内的错误不会向外抛出，一般通过 try catch 解决。在这种情况下需要注意，处理错误后，仍然要抛出运行时异常，以触发事务的回滚。

3．事务配置的传播方式

当事务的传播方式配置为 SUPPORTS 时，在当前没有事务的情况下，不会启用事务。此错误在日常开发中注意即可。

4．数据库不支持事务

此情况发生的频率不高，需要注意的是，当使用 MySQL 数据库时，数据库引擎为 MyISAM 时是不支持事务的。

6.5 本章小结

本章首先介绍了事务的管理方式，然后分别讲解了基于注解与 XML 实现事务管理的方法，最后讲解了事务的传播方式。使用注解方式完成声明式事务的管理是本章的重点。

在本章内容中，事务被分为两种，一种是声明式事务，另一种是编程式事务。编程式事务在日常开发中不常使用，读者可以初步了解。声明式事务是重点内容，在声明式事务中，使用注解方式完成的声明式事务占实际开发中的绝大部分。此外，事务的传播方式是面试时常问的知识点，读者可按需掌握。

6.6 习题

1．填空题

（1）在 Spring 中，负责处理事务的注解是________。

（2）事务的管理方式包括________事务管理和________事务管理等。

（3）在@Transaction 注解的属性中，负责设置事务隔离级别的是________。

2．选择题

（1）关于事务，下列描述错误的是（　　）。

A. 事务是基于 AOP 实现的

B. 事务默认的传播方式是 PROPAGATION_REQUIRES_NEW

C. 在同一个类中，非事务方法调用事务方法会导致事务失效

D. 事务中发生异常，但此异常被捕获时，此事务不会回滚

（2）关于事务的传播方式，下列叙述错误的是（　　）。

A. 当传播方式为 PROPAGATION_MANDATORY 时，如果不存在事务，则报错

B. 当传播方式为 PROPAGATION_REQUIRES_NEW 时，如果存在事务，则仍然新建事务，在新建的事务中运行

C. 当传播方式为 PROPAGATION_NOT_SUPPORTED 时，如果存在事务，则报错

D. 当传播方式为 PROPAGATION_REQUIRED 时，如果不存在事务，则新建一个事务

（3）关于@Transactional 注解的属性，下列描述错误的是（　　）。

A. noRollbackFor 属性用于指定事务遇到特定异常时不回滚

B. propagation 属性负责指定事务的传播方式

C. readOnly 属性用于指定事务是否为只读

D. value 属性用于指定事务的名称

3．思考题

（1）简述事务的传播方式。

（2）简述事务失效的原因。

4．编程题

创建 Account 类，在其中添加 number 属性，并创建对应数据库。创建 Operation 类，在其中添加 trade()方法，在此方法中使用 JdbcTemplate 对 Account 类进行两次操作，一次增加数据操作和一次减少数据操作，在这两个操作之间模拟空指针异常，随后使用事务对其进行处理（使用注解方式）。

第7章 MyBatis 基础

本章学习目标

- 了解 ORM 框架的原理。
- 了解 MyBatis 的工作流程和核心对象。
- 理解 MyBatis 的基本概念。
- 掌握 MyBatis 的下载方法。
- 掌握 MyBatis 入门程序的编写。

MyBatis 基础

Java 程序依靠 JDBC 实现对数据库的操作。从第 4 章可以了解到，Spring 提供了 JdbcTemplate 来简化数据库操作，但是对 JDBC 模板的操作仍然有些复杂。在实际开发中，JdbcTemplate 的复用性差，在每次连接数据库时，都会引入不必要的类。随着 ORM 框架的兴起，以 ORM 框架为基础的 MyBatis 框架逐渐得到重视，并广泛应用于企业级开发中。本章将对 MyBatis 框架涉及的基础知识进行讲解。

7.1 MyBatis 概述

2001 年，Clinton Begin 发起开源项目 iBatis。2010 年，iBatis 由 Apache Software Foundation 迁移到 Google Code，并改名为 MyBatis。2013 年 11 月，MyBatis 迁移到 GitHub，由 GitHub 维护。本节将讲解 MyBatis 的基本原理，读者可初步了解 ORM 框架的原理，理解 MyBatis 的基本概念。

7.1.1 ORM 框架

在当今企业级应用的开发中，对象和关系数据是业务实体的两种表现形式。业务实体在内存中表现为对象，在数据库中表现为关系数据。当采用面向对象的方法编写程序时，一旦需要访问数据库，就需要采用关系数据的访问方式，这种转换为开发人员带来了很大的麻烦。

ORM 框架是一个对象-关系映射的系统化解决方案，当 ORM 框架完成转换后，开发人员可以直接取用对象。常用的 ORM 框架有 Hibernate 和 MyBatis，这两者之间的区别在 1.2 节中进行了详细介绍。

ORM 框架将数据库查询到的数据封装为实体类对象，ORM 映射流程如图 7.1 所示。

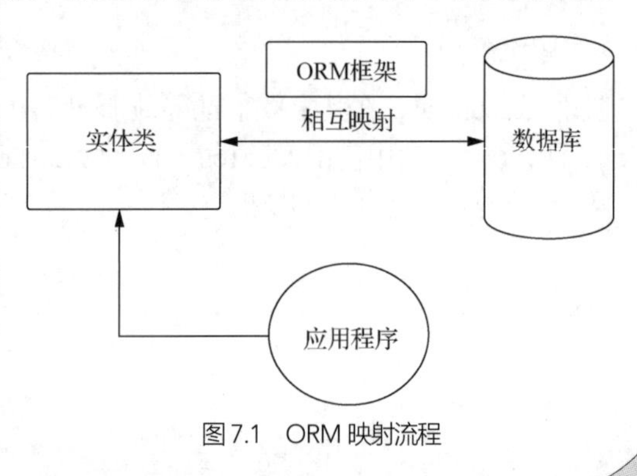

图 7.1　ORM 映射流程

从图 7.1 中可以看出，实体类与数据库之间通过 ORM 框架相互映射，应用程序可以直接获取映射完成的实体类。

7.1.2 MyBatis 简介

MyBatis 是一款优秀的 ORM 框架，它需要手动编写 POJO、SQL 语句并匹配映射关系，因此它可以更加灵活地生成映射关系。MyBatis 允许开发人员利用数据库的各项功能完成复杂的操作，例如存储过程、视图或进行复杂查询等，它具有高度灵活、可优化和易维护等优点。

此外，MyBatis 支持动态列、动态表和存储过程，同时提供了简易的日志、缓存和级联功能，是企业级开发强有力的框架。

7.2 MyBatis 的工作流程和核心对象

本节将讲解 MyBatis 的工作流程和核心对象。读者需要深入了解 MyBatis 的工作流程，熟练掌握 SqlSessionFactory 和 SqlSession 等核心对象的使用方法。

7.2.1 工作流程

MyBatis 的工作流程是 MyBatis 中重要的知识点，如图 7.2 所示。

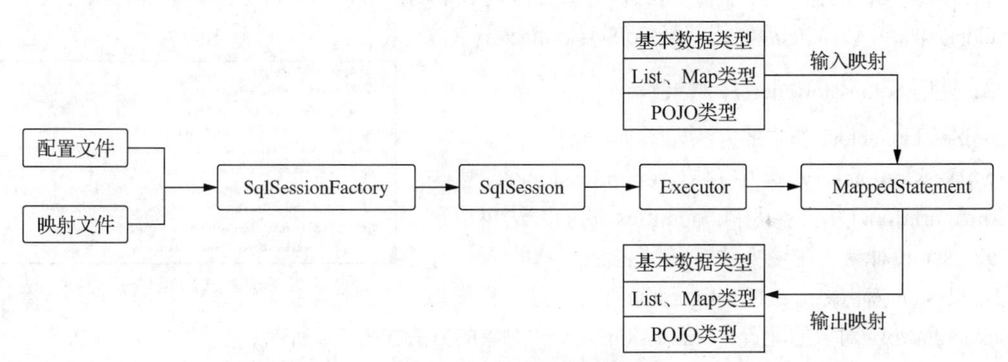

图 7.2　MyBatis 的工作流程

在图 7.2 中，MyBatis 的工作流程分为如下 5 个步骤。

（1）编写配置文件与映射文件，其中，配置文件用于进行数据库连接的设置，映射文件用于进行与 SQL 文件相关的设置。

（2）MyBatis 通过配置文件和映射文件生成 SqlSessionFactory 对象，此对象在 MyBatis 的整个生命周期中只存在一个，它负责为每一次连接生成 SqlSession 对象。

（3）通过 SqlSessionFactory 对象生成 SqlSession 对象，SqlSession 对象在每次连接中只有一个，它封装了操作数据库的所有方法。

（4）在每一次连接中，通过 SqlSession 对象操作数据库，SqlSession 对象通过底层的 Executor 执行器执行对应操作。Executor 执行器有两种，一种是普通执行器，另一种是缓存执行器，。

（5）Executor 执行器将操作封装为 MappedStatement 对象，在执行 SQL 语句之前，Executor 执行器通过 MappedStatement 对象将输入的实体对象或基本类型数据映射到 SQL 语句。在执行 SQL 语句之后，Executor 执行器通过 MappedStatement 对象将 SQL 语句的执行结果映射为实体对象或基本类型数据。

7.2.2 SqlSessionFactory 与 SqlSession

SqlSessionFactory 与 SqlSession 是 MyBatis 中的核心对象，下面对它们的使用方法进行讲解。

1．创建 SqlSessionFactory 对象

创建 SqlSessionFactory 对象需要使用 SqlSessionFactoryBuilder 类，SqlSessionFactoryBuilder 类中的方法如图 7.3 所示。

调用 SqlSessionFactoryBuilder 类中的 build()方法可创建 SqlSessionFactory 对象。从图 7.3 中可以看出，build()有多种重载方法，其参数可以选填 Reader 和 InputStream 的实现类。在开发中，经常使用的 build()方法如下所示。

```
build(InputStream inputStream):SqlSessoinFactory
```

SqlSessionFactoryBuilder
SqlSessionFactoryBuilder()
build(Reader): SqlSessionFactory
build(Reader, String): SqlSessionFactory
build(Reader, Properties): SqlSessionFactory
build(Reader, String, Properties): SqlSessionFactory
build(InputStream): SqlSessionFactory
build(InputStream, String): SqlSessionFactory
build(InputStream, Properties): SqlSessionFactory
build(InputStream, String, Properties): SqlSessionFactory
build(Configuration): SqlSessionFactory

图 7.3　SqlSessionFactoryBuilder 类中的方法

build()方法的参数需要 MyBatis 配置文件的输入流，下面创建 MyBatis 配置文件的输入流。调用 Resources 类的 getResourceAsStream()方法，传入配置文件的绝对路径，如下所示。

```
InputStream inputStream = Resources.getResourceAsStream("文件路径");
```

获得配置文件的输入流之后，将其作为参数传入 SqlSessionFactoryBuilder 类的 build()方法中，调用 build()方法，该方法的返回值就是 SqlSessionFactory 对象。

2．使用 SqlSessionFactory 对象

SqlSessionFactory 类中的方法如图 7.4 所示。

SqlSessionFactory
openSession(): SqlSession
openSession(boolean): SqlSession
openSession(Connection): SqlSession
openSession(TransactionIsolationLevel): SqlSession
openSession(ExecutorType): SqlSession
openSession(ExecutorType, boolean): SqlSession
openSession(ExecutorType, TransactionIsolationLevel): SqlSession
openSession(ExecutorType, Connection): SqlSession
getConfiguration(): Configuration

图 7.4　SqlSessionFactory 类中的方法

SqlSessionFactory 类中存在 openSession()方法与 getConfiguration()方法。其中，openSession()方法可以创建 SqlSession 对象，在该方法中传入参数可以设置创建的 SqlSession 对象；getConfiguration()方法用于获取 SqlSessionFactory 对象的配置。SqlSessionFactory 对象的方法如表 7.1 所示。

表 7.1　SqlSessionFactory 对象的方法

方法名称	说明
SqlSession openSession()	开启一个连接，开启后事务的传播方式将使用默认设置
SqlSession openSession(Boolean autoCommit)	参数 autoCommit 用于设置是否自动提交
SqlSession openSession(Connection connection)	使用连接的参数进行配置
SqlSession openSession(TransactionIsolationLevel level)	指定连接事务的传播方式
SqlSession openSession(ExecutorType executorType)	指定执行器类型
SqlSession openSession(ExecutorType executorType, Boolean autoCommit)	指定执行器类型和是否自动提交
SqlSession openSession(ExecutorType executorType, Connection connection)	指定执行器类型和连接
Configuration getConfiguration()	获取 Configuration 对象，此对象代表 SqlSessionFactory 的全部配置

3．使用 SqlSession 对象

SqlSession 对象是 MyBatis 中的核心对象。在日常开发中，常用 SqlSession 对象与数据库进行交互。除此之外，SqlSession 对象贯穿整个数据库访问的过程，若一定时间段内不使用 SqlSession 对象，需要及时调用 SqlSession 对象的 close()方法将其关闭。

SqlSession 对象提供了执行 SQL 语句、提交事务、回滚事务、使用映射器等方法，在方法中需要指定映射文件中的方法。SqlSessionFactory 对象的方法如表 7.2 所示。

表 7.2　SqlSession 对象的方法

方法名称	说明
T selectOne(String var1)	执行单条记录的查询操作，需传入执行查询的方法，返回映射的对象
T selectOne(String var1, Object var2)	执行单条记录的查询操作，需传入查询所需的方法和参数，返回映射的对象
List<E> selectList(String var1)	执行多条记录的查询操作，需传入查询方法，返回查询结果的集合
List<E> selectList(String var1, Object var2)	执行多条记录的查询操作，需传入查询方法和参数，返回查询结果的集合
Map<K, V> selectMap(String var1, String var2)	执行查询操作，返回一个映射查询结果的 Map 集合
Map<K, V> selectMap(String var1, Object var2, String var3)	执行查询操作，需传入查询的方法和参数，返回 Map 集合
int insert(String var1)	执行插入操作，需传入映射文件中的方法名，返回数据库中受影响的数据行数
int insert(String var1, Object var2)	执行插入操作，需传入映射文件中的方法名和参数对象，返回数据库中受影响的数据行数
int update(String var1)	执行更新操作，需传入映射文件中的方法名，返回数据库中受影响的数据行数
int update(String var1, Object var2)	执行更新操作，需传入映射文件中的方法名和参数对象，返回数据库中受影响的数据行数
int delete(String var1)	执行删除操作，需传入映射文件中的方法名，返回数据库中受影响的数据行数
int delete(String var1, Object var2)	执行参数操作，需传入映射文件中的方法名和参数对象，返回数据库中受影响的数据行数
commit()	提交事务
commit(boolean var1)	Var1 默认为 false，参数值为 true 时表示强制提交
rollback()	回滚
rollback(boolean var1)	强制回滚
close()	关闭 SqlSession 对象
T getMapper(Class<T> var1)	获取映射器

7.3 MyBatis 应用示例

本节将以一个完整的示例讲解 MyBatis 的下载与应用，读者需要熟练掌握本节全部内容。

7.3.1 MyBatis 的下载

MyBatis 为第三方组件，不属于 Spring，也不依赖 Spring。因此，在编写代码之前，需要下载 MyBatis

的依赖包。

（1）访问 GitHub 官网提供的 MyBatis 下载地址，如图 7.5 所示。

（2）单击图 7.5 中的 mybatis-3.5.7.zip，下载文件。

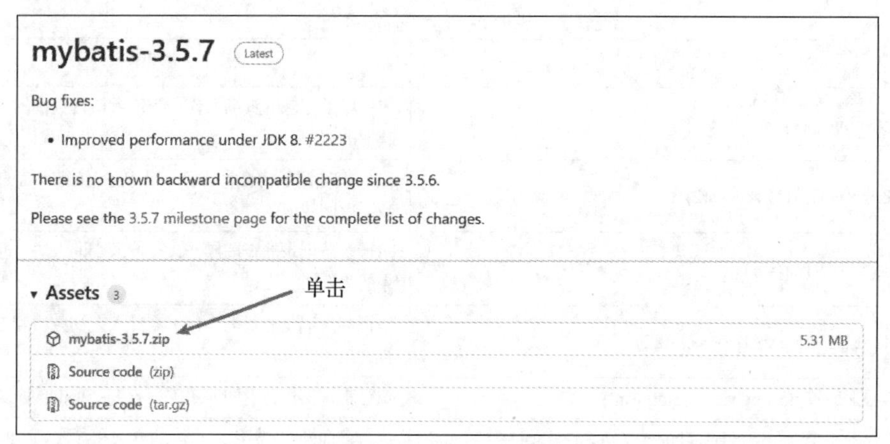

图 7.5　下载 MyBatis 依赖包

（3）解压下载后的文件，其目录结构如图 7.6 所示。

（4）将 mybatis-3.5.7.jar 和 lib 文件夹下所有的依赖包复制到 IDEA 的 lib 文件夹中，然后将复制的依赖包添加到工程中即可。

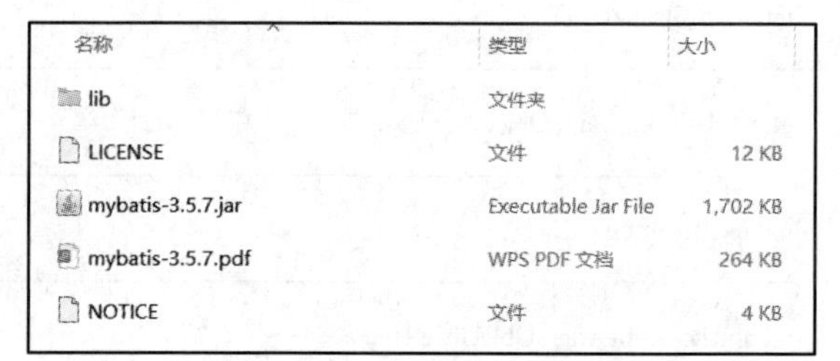

图 7.6　MyBatis 依赖包的目录结构

7.3.2　MyBatis 的简单应用

下面通过一个完整的示例讲解 MyBatis 的使用方法，此示例中的代码可以在非 Spring 环境中运行。MyBatis 与 Spring 之间整合的方法将在第 12 章讲解。

1．创建数据表

在 MySQL 中创建 dog 表，代码如例 7-1 所示。

【例 7-1】 dog.sql

```
1.  DROP TABLE IF EXISTS 'dog';
2.  CREATE TABLE 'dog'  (
3.    'id' int(0) NOT NULL AUTO_INCREMENT,
4.    'name' varchar(255) CHARACTER,
5.    'age' int(0) NULL DEFAULT NULL,
6.    PRIMARY KEY ('id') USING BTREE
7.  )
```

在 dog 表中插入数据，代码如例 7-2 所示。

【例 7-2】 dog_insert.sql

```
1.  INSERT INTO 'dog' VALUES (1, '小红', 4);
2.  INSERT INTO 'dog' VALUES (2, '旺财', 2);
3.  INSERT INTO 'dog' VALUES (3, '胖胖', 3);
```

查看数据表的结构与数据，如图 7.7 所示。

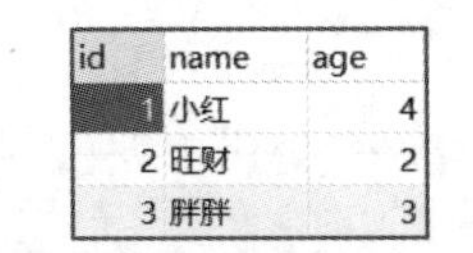

id	name	age
1	小红	4
2	旺财	2
3	胖胖	3

图 7.7　dog 表的结构与数据

从图 7.7 中可以看出，表创建成功，数据插入成功。

2．创建实体类

MyBatis 框架将数据库的查询结果映射到实体类中，因此，需要创建一个查询结果的映射类，在 src 文件夹下创建 com 文件夹，然后在 com 文件夹下创建 Dog 类（此处的代码中省略 set()方法、get()方法、toString()方法、全参构造方法和无参构造方法，在实际应用中需要添加），代码如例 7-3 所示。

【例 7-3】 Dog.java

```
1.  package com;
2.
3.  public class Dog {
4.      Integer id;
5.      String name;
6.      Integer age;
7.  }
```

3．创建配置文件

MyBatis 需要配置文件来加载配置，在 src 文件夹下创建 mybatis-config.xml 文件，其中的代码如例 7-4 所示。

【例 7-4】 mybatis-config.xml

```
1.  <?xml version="1.0" encoding="UTF-8" ?>
2.  <!DOCTYPE configuration
3.          PUBLIC "-//mybatis.org//DTD Config 3.0//EN"
4.          "http://mybatis.org/dtd/mybatis-3-config.dtd">
5.  <configuration>
6.
7.      <!--配置环境-->
8.      <environments default="mysql">
9.          <!--配置 MySQL 环境-->
10.         <environment id="mysql">
11.             <!--配置事务管理器-->
12.             <transactionManager type="JDBC"/>
13.             <!--配置数据库连接-->
14.             <dataSource type="POOLED">
15.                 <!--配置数据库连接驱动-->
16.                 <property name="driver" value="com.mysql.cj.jdbc.Driver"/>
17.                 <!--配置数据库连接地址-->
18.                 <property name="url"
19.                         value="jdbc:mysql://127.0.0.1/test?
20.                                 characterEncoding=utf8&amp;
21.                                 useSSL=false&amp;
22.                                 serverTimezone=UTC&amp;
23.                                 allowPublicKeyRetrieval=true"/>
24.                 <!--配置用户名-->
25.                 <property name="username" value="root"/>
```

```
26.             <!--配置密码-->
27.             <property name="password" value="******"/>
28.         </dataSource>
29.     </environment>
30. </environments>
31.
32. </configuration>
```

在例 7-4 中，利用<environment>元素配置 MySQL 环境，利用<transactionManager>元素配置事务管理器，利用<dataSource>元素配置数据库连接。下面创建映射文件，并将映射文件配置到 MyBatis 的配置文件中。

4．创建映射文件

为了方便文件的管理，在 com 文件夹下创建 mapper 文件夹，在 mapper 文件夹下创建 Dog 类对应的映射文件 DogMapper.xml。此示例的目录结构如图 7.8 所示。

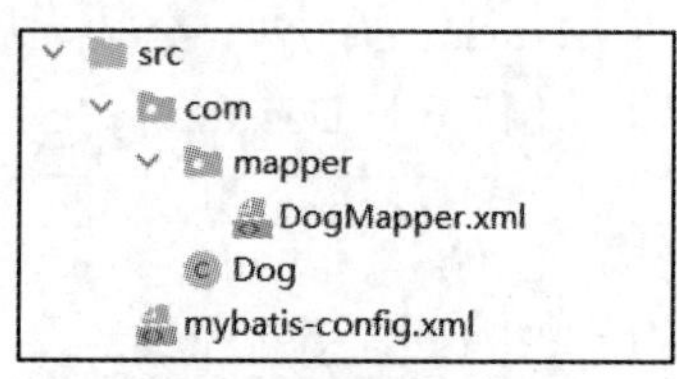

图 7.8　示例的目录结构

编写 DogMapper.xml 文件，代码如例 7-5 所示。

【例 7-5】 DogMapper.xml

```
1.  <?xml version="1.0" encoding="UTF-8"?>
2.  <!DOCTYPE mapper PUBLIC "-//mybatis.org//DTD Mapper 3.0//EN"
3.          "http://mybatis.org/dtd/mybatis-3-mapper.dtd">
4.  <mapper namespace="student">
5.      <select id="selectDog" resultType="com.Dog">
6.          select * from dog
7.      </select>
8.  </mapper>
```

下面介绍例 7-5 中的元素。

（1）<mapper>元素是 MyBatis 映射文件的根元素，元素中的 namespace 属性代表映射文件的名称。

（2）在<mapper>元素中可以定义操作元素，<select>元素为操作元素的一种，负责查询数据库中的数据。在<select>元素中，id 属性是操作的名称，resultType 属性代表查询结果的映射对象。<select>元素的内部可以添加查询操作的 SQL 语句，例 7-5 中执行的是 dog 表的全查询，查询之后将结果映射到 Dog 类中，并将 Dog 类返回。

在完成映射文件的编写之后，需要将此文件配置到 MyBatis 的配置文件中，相关代码如例 7-6 所示。

【例 7-6】 mapperInsert.xml

```
1.  <mappers>
2.          <mapper resource="com/mapper/DogMapper.xml"/>
3.  </mappers>
```

例 7-6 中的代码需要添加到 mybatis-config.xml 文件的<environments>元素中。在例 7-6 中，<mappers>元素负责配置映射文件，其内部可以配置多个<mapper>元素。每个<mapper>元素代表一个映射文件，其中的 resource 属性负责指定映射文件的路径。

5．测试数据库操作

在 com 文件夹下创建测试类，代码如例 7-7 所示。

【例 7-7】 Test1.java

```
1.  public class Test {
2.      public static void main(String[] args) {
```

```
3.
4.          /*创建输入流*/
5.          InputStream inputStream = null;
6.
7.          /*将MyBatis 配置文件转换为输入流*/
8.          try {
9.             inputStream =
10.                    Resources.getResourceAsStream("mybatis-config.xml");
11.         } catch (IOException e) {
12.            e.printStackTrace();
13.         }
14.
15.         /*通过 SqlSessionFactoryBuilder 类创建 SqlSessionFactory 对象*/
16.         SqlSessionFactory build =
17.                new SqlSessionFactoryBuilder().build(inputStream);
18.
19.         /*通过 SqlSessionFactory 类创建 SqlSession 对象*/
20.         SqlSession sqlSession = build.openSession();
21.
22.      /*通过 SqlSession 类的 selectList()方法执行 student 映射文件中的 selectDog 操作*/
23.         List<Object> objects = sqlSession.selectList("student.selectDog");
24.
25.         /*将返回值输出*/
26.         for (Object object : objects) {
27.            System.out.println((Dog)object);
28.         }
29.         /*关闭事务*/
30.         sqlSession.close();
31.      }
32. }
```

下面介绍例 7-7 中的代码。

（1）在代码的第 5 行中创建输入流，并将其赋值为空。

（2）在代码的第 10 行中，调用 Resources 类的 getResourceAsStream()方法将 MyBatis 配置文件转换为输入流。

（3）在代码的第 17 行中，调用 SqlSessionFactoryBuilder 类的 build()方法，在 build()方法中传入 MyBatis 配置文件的输入流，获取返回的 SqlSessionFactory 对象。

（4）在代码的第 20 行中，调用 SqlSessionFactory 类的 openSession()方法，获取返回的 SqlSession 对象。

（5）在代码的第 23 行中，调用 SqlSession 类的 selectList()方法，在 selectList()方法中传入需要执行的操作名称，获取返回的查询结果集。

（6）在代码的第 27 行中将返回值输出。

运行测试类，结果如图 7.9 所示。

从图 7.9 可以看出，输出的结果为 dog 表中的所有数据，测试成功。

```
Dog{id=1, name='小红', age=4}
Dog{id=2, name='旺财', age=2}
Dog{id=3, name='胖胖', age=3}

Process finished with exit code 0
```

图 7.9 运行结果

7.3.3 SqlSession 的增删改查操作

在例 7-7 中使用了 SqlSession 类的 selectList()方法，通过该方法执行多条查询操作的语句，将查询的结果封装为集合。下面在 7.3.2 小节的基础上，测试 SqlSession 类的增删改查方法。

1. SqlSession 的新增操作

在 DogMapper.xml 文件的<mapper>元素中加入映射信息，具体代码如下。

```
<insert id="insertDog" parameterType="com.Dog">
      insert into dog(name,age) value(#{name},#{age})
</insert>
```

在以上代码中，利用<insert>元素创建一个插入操作，其中，parameterType 属性用于指定传入的参数对象类型，可以省略，此处传入的是 Dog 对象。在<insert>元素内使用“#{}”方式取用参数对象的属性，在此取用的是 Dog 对象的 name 和 age 属性。

在 com 文件夹下创建 TestInsert 类，具体代码如例 7-8 所示。

【例 7-8】 TestInsert.java

```
1.  public class TestInsert {
2.      public static void main(String[] args) {
3.
4.          /*创建输入流*/
5.          InputStream inputStream = null;
6.
7.          /*将 MyBatis 配置文件转换为输入流*/
8.          try {
9.              inputStream =
10.                     Resources.getResourceAsStream("mybatis-config.xml");
11.         } catch (IOException e) {
12.             e.printStackTrace();
13.         }
14.
15.         /*通过 SqlSessionFactoryBuilder 类创建 SqlSessionFactory 对象*/
16.         SqlSessionFactory build =
17.                 new SqlSessionFactoryBuilder().build(inputStream);
18.
19.         /*通过 SqlSessionFactory 类创建 SqlSession 对象*/
20.         SqlSession sqlSession = build.openSession();
21.
22.         /*创建对象*/
23.         Dog dog = new Dog();
24.         dog.setName("旺财二号");
25.         dog.setAge(3);
26.
27.         /*使用对象传参*/
28.         int stateObject = sqlSession.insert("student.insertDog", dog);
29.
30.         HashMap mapper = new HashMap();
31.         mapper.put("name","旺财三号");
32.         mapper.put("age",3);
33.
34.         /*使用对象传参*/
35.         int stateMap = sqlSession.insert("student.insertDog", mapper);
36.
37.         System.out.println(stateObject+"-"+stateMap);
```

```
38.
39.         /*提交事务*/
40.         sqlSession.commit();
41.         /*关闭事务*/
42.         sqlSession.close();
43.     }
44. }
```

在代码的第 28 行中，调用 SqlSession 类的 insert()方法。此方法中有两个参数，第一个参数是传入的方法名，第二个参数是传入的参数对象。

在代码的第 35 行中，调用 SqlSession 的 insert()方法传入 Map 对象，当第二个参数的值不能用对象封装时，可以使用 Map 集合来代替对象。值得注意的是，Map 集合的键要与“#{}”中的名字保持一致，并且<insert>元素中的 parameterType 属性值需要更换成 java.util.Map。

在代码的第 40 行中，调用 SqlSession 类的 commit()方法提交事务。

在代码的第 42 行中，调用 SqlSession 类的 close()方法关闭事务。

执行 TestInsert 类，输出结果如下所示。

```
1-1
```

输出结果为 1-1，表示两次操作均成功。

2．SqlSession 的修改操作

在 DogMapper.xml 文件的<mapper>元素中加入映射信息，具体代码如下所示。

```
<update id="updateDog" parameterType="java.util.Map">
    update dog set age = #{age} where id=1
</update>
```

在以上代码中，利用<update>元素创建一个更新操作，将 parameterType 属性设置为 Map 类型。在<update >元素内使用“#{}”方式取用 Map 集合中的属性。

在 com 文件夹下创建 TestUpdate 类，具体代码如例 7-9 所示。

【例 7-9】 TestUpdate.java

```
1.  public class TestUpdate {
2.      public static void main(String[] args) {
3.
4.          /*创建输入流*/
5.          InputStream inputStream = null;
6.
7.          /*将 MyBatis 配置文件转换为输入流*/
8.          try {
9.              inputStream =
10.                     Resources.getResourceAsStream("mybatis-config.xml");
11.         } catch (IOException e) {
12.             e.printStackTrace();
13.         }
14.
15.         /*通过 SqlSessionFactoryBuilder 类创建 SqlSessionFactory 对象*/
16.         SqlSessionFactory build =
17.                 new SqlSessionFactoryBuilder().build(inputStream);
18.
19.         /*通过 SqlSessionFactory 类创建 SqlSession 对象*/
```

```
20.         SqlSession sqlSession = build.openSession();
21.
22.         HashMap mapper = new HashMap();
23.         mapper.put("age",3);
24.
25.         /*使用 Map 集合传参*/
26.         int stateMap = sqlSession.update("student.updateDog", mapper);
27.
28.         System.out.println(stateMap);
29.
30.         /*提交事务*/
31.         sqlSession.commit();
32.         /*关闭事务*/
33.         sqlSession.close();
34.     }
35. }
```

在代码的第 26 行中，调用 SqlSession 类的 update()方法，将需要的类与参数传入，返回值为此操作影响的数据行数。执行 TestUpdate 类，输出结果如下所示。

```
1
```

输出结果为 1，表示操作成功。

3．SqlSession 的删除操作

在 DogMapper.xml 文件的<mapper>元素中加入映射信息，具体代码如下所示。

```
<delete id="deleteDog">
    delete from dog where id = #{id}
</delete>
```

在以上代码中，利用<delete>元素创建一个删除操作，将 parameterType 属性忽略，在<update>元素内使用“#{}”方式取用属性。

在 com 文件夹下创建 TestDelete 类，具体代码如例 7-10 所示。

【例 7-10】 TestDelete.java

```
1.  public class TestDelete {
2.      public static void main(String[] args) {
3.
4.          /*创建输入流*/
5.          InputStream inputStream = null;
6.
7.          /*将 MyBatis 配置文件转换为输入流*/
8.          try {
9.              inputStream =
10.                     Resources.getResourceAsStream("mybatis-config.xml");
11.         } catch (IOException e) {
12.             e.printStackTrace();
13.         }
14.
15.         /*通过 SqlSessionFactoryBuilder 类创建 SqlSessionFactory 对象*/
16.         SqlSessionFactory build =
17.                 new SqlSessionFactoryBuilder().build(inputStream);
18.
```

```
19.         /*通过 SqlSessionFactory 类创建 SqlSession 对象*/
20.         SqlSession sqlSession = build.openSession();
21.
22.         int stateMap = sqlSession.delete("student.deleteDog", 3);
23.
24.         System.out.println(stateMap);
25.
26.         /*提交事务*/
27.         sqlSession.commit();
28.         /*关闭事务*/
29.         sqlSession.close();
30.     }
31. }
```

在代码的第22行中，调用 SqlSession 类的 delete()方法，将需要的类与参数传入。需要注意的是，当该方法中传入的参数唯一时，可以不为此参数设置名称；在通过“#{}”形式取用此参数时，大括号中的内容无论是什么，最终获取到的都是传入的唯一参数。

执行 TestUpdate 类，输出结果如下所示。

```
1
```

输出结果为1，表示操作成功。

4．SqlSession 的模糊查询操作

在 DogMapper.xml 文件的<mapper>元素中加入映射信息，具体代码如下所示。

```
<select id="selectByName" resultType="com.Dog">
    select * from dog where name like '%${name}%'
</select>
```

在以上代码中，利用<select>元素创建一个查询操作，resultType 属性用于指定返回值类型，此处的返回值类型为 Dog 类，在<select>元素内使用“${}”方式取用属性。

“#{}”表示占位符，在拼接 SQL 语句的过程中，先用占位符进行占位，然后将取得的值放到此位置，值的类型有严格的限制。“${}”表示拼接符，在拼接 SQL 语句的过程中，此符号用于将取得的值直接拼接到相应位置。

此映射信息需要的参数为 name，需要使用字符串拼接，因此使用“${}”。

在 com 文件夹下创建 TestSelectByName 类，具体代码如例 7-11 所示。

【例 7-11】 TestSelectByName.java

```
1.  public class TestSelectByName {
2.      public static void main(String[] args) {
3.
4.          /*创建输入流*/
5.          InputStream inputStream = null;
6.
7.          /*将 MyBatis 配置文件转换为输入流*/
8.          try {
9.              inputStream =
10.                     Resources.getResourceAsStream("mybatis-config.xml");
11.         } catch (IOException e) {
12.             e.printStackTrace();
13.         }
```

```
14.
15.         /*通过 SqlSessionFactoryBuilder 类创建 SqlSessionFactory 对象*/
16.         SqlSessionFactory build =
17.               new SqlSessionFactoryBuilder().build(inputStream);
18.
19.         /*通过 SqlSessionFactory 类创建 SqlSession 对象*/
20.         SqlSession sqlSession = build.openSession();
21.
22.         HashMap mapper = new HashMap();
23.         mapper.put("age",3);
24.
25.         List<Object> result =
26.                 sqlSession.selectList("student.selectByName", "旺财");
27.
28.         /*将返回值输出*/
29.         for (Object object : result) {
30.            System.out.println((Dog)object);
31.         }
32.
33.         /*关闭事务*/
34.         sqlSession.close();
35.     }
36. }
```

在代码的第 26 行中，调用 SqlSession 类的 selectList()方法，传入方法名称与参数，其返回值使用集合进行接收。

在代码的第 30 行中，将查询到的结果输出，具体如图 7.10 所示。

从图 7.10 可以看出，输出结果是 name 属性中含有“旺财”的全部记录。

```
Dog{id=2, name='旺财', age=2}
Dog{id=13, name='旺财二号', age=3}
Dog{id=14, name='旺财三号', age=3}

Process finished with exit code 0
```

图 7.10 输出结果

7.4 MyBatis 接口开发

使用 XML 文件进行开发，在调用 SqlSession 进行操作时，需要指定 MyBatis 配置文件中的方法，这种调用方式过于烦琐。为解决此问题，MyBatis 提供了接口开发的方式，本节将通过示例详细讲解接口开发的方式。

1. 创建目录

创建规范的目录结构。在工程文件夹下创建 resource 文件夹，右键单击 resource 文件夹，在弹出的快捷菜单中选择 Mark Directory as →Resources Root 命令，将其标记为资源文件夹，如图 7.11 所示（此后标记资源文件夹的操作不再讲解）。

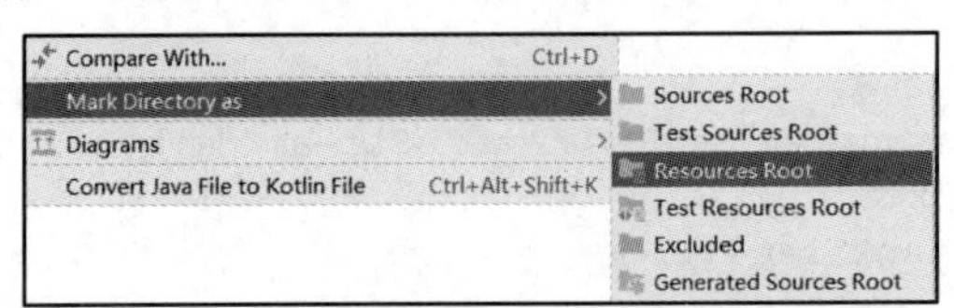

图 7.11 标记资源文件夹

在 src 文件夹下创建 com 文件夹，在 com 文件夹下创建 mapper 文件夹和 pojo 文件夹，至此，目录结构创建完毕。在 pojo 文件夹下存放实体类文件，在 mapper 文件夹下存放 XML 映射文件与 Mapper 接口文件，在 resource 文件夹下存放配置文件。

2．创建实体类与数据表

在 pojo 文件夹下创建 Dog 类，示例采用例 7-3。在数据库中创建 dog 表，创建 dog 表的 SQL 语句如例 7-1 所示。

3．创建配置文件

在 resource 文件夹下创建 MyBatis 配置文件，配置文件中的代码如例 7-12 所示。

【例 7-12】 mybatis-config.xml

```
1.  <?xml version="1.0" encoding="UTF-8" ?>
2.  <!DOCTYPE configuration
3.         PUBLIC "-//mybatis.org//DTD Config 3.0//EN"
4.         "http://mybatis.org/dtd/mybatis-3-config.dtd">
5.  <configuration>
6.
7.      <!--配置环境-->
8.      <environments default="mysql">
9.          <!--配置 MySQL 环境-->
10.         <environment id="mysql">
11.             <!--配置事务管理器-->
12.             <transactionManager type="JDBC"/>
13.             <!--配置数据库连接-->
14.             <dataSource type="POOLED">
15.                 <!--配置数据库连接驱动-->
16.                 <property name="driver" value="com.mysql.cj.jdbc.Driver"/>
17.                 <!--配置数据库连接地址-->
18.                 <property name="url"
19.                         value="jdbc:mysql://127.0.0.1/test?
20.                                characterEncoding=utf8&amp;
21.                                useSSL=false&amp;
22.                                serverTimezone=UTC&amp;
23.                                allowPublicKeyRetrieval=true"/>
24.                 <!--配置用户名-->
25.                 <property name="username" value="root"/>
26.                 <!--配置密码-->
27.                 <property name="password" value="******"/>
28.             </dataSource>
29.         </environment>
30.     </environments>
31.     <mappers>
32.         <!--将 com.mapper 包下的所有 Mapper 接口引入-->
33.         <package name="com.mapper"/>
34.     </mappers>
35. </configuration>
```

与 7.3 节的 MyBatis 配置文件相比，此文件中<mappers>元素中使用的是<package>元素，<package>元素的 name 属性负责指定 Mapper 接口文件的包名，MyBatis 在运行时会将此包中的所有接口实现，并为此实现类添加方法。其中，每个方法对应映射文件中的一个操作。

4．创建映射文件与 Mapper 接口文件

在 mapper 文件夹下创建 DogMapper 文件，代码如例 7-13 所示。

【例 7-13】 DogMapper.java

```
1.  public interface DogMapper {
2.      /*查询*/
3.      Dog selectDog();
4.  }
```

在例 7-13 中，接口文件的名称为 DogMapper，此名与映射文件名需要保持一致，即映射文件名为 DogMapper.xml。此外，DogMapper 接口中的方法名应与映射文件中操作的 id 值保持一致。DogMapper.xml 对应的映射文件的代码如例 7-14 所示。

【例 7-14】 DogMapper.xml

```
1.  <?xml version="1.0" encoding="UTF-8"?>
2.  <!DOCTYPE mapper PUBLIC "-//mybatis.org//DTD Mapper 3.0//EN"
3.          "http://mybatis.org/dtd/mybatis-3-mapper.dtd">
4.  <mapper namespace="com.mapper.DogMapper">
5.      <select id="selectDog" resultType="dog">
6.          select * from dog where id = 1
7.      </select>
8.  </mapper>
```

在例 7-14 中，<mapper>元素的 namespace 属性需要赋值为对应 Mapper 接口的绝对路径。此外，<select>元素中的 id 属性值需要和 Mapper 接口中的方法名对应。

5．创建测试类

在 com 文件夹下创建测试类，代码如例 7-15 所示。

【例 7-15】 Test.java

```
1.  public class Test {
2.      public static void main(String[] args) {
3.          /*创建输入流*/
4.          InputStream inputStream = null;
5.
6.          /*将 MyBatis 配置文件转换为输入流*/
7.          try {
8.              inputStream =
9.                      Resources.getResourceAsStream("mybatis-config.xml");
10.         } catch (IOException e) {
11.             e.printStackTrace();
12.         }
13.
14.         /*通过 SqlSessionFactoryBuilder 类创建 SqlSessionFactory 对象*/
15.         SqlSessionFactory build =
16.                 new SqlSessionFactoryBuilder().build(inputStream);
17.
18.         /*通过 SqlSessionFactory 类创建 SqlSession 对象*/
19.         SqlSession sqlSession = build.openSession();
20.
21.         DogMapper mapper = sqlSession.getMapper(DogMapper.class);
22.         Dog dog1 = mapper.selectDog();
23.
```

```
24.         System.out.println(dog1);
25.
26.         /*关闭事务*/
27.         sqlSession.close();
28.     }
29. }
```

例 7-15 中的第 21 行代码，通过调用 SqlSession 类的 getMapper()方法获取 MyBatis 提供的 Mapper 接口的实现类，并传入 Mapper 接口实现类的 class 对象作为参数。当获取到实现类后，就可以调用其中的方法。

测试类的运行结果如下所示。

```
Dog{id=1, name='小红', age=3}
```

该项目的整体目录结构如图 7.12 所示。

图 7.12 所示为开发中常用的目录结构，读者需要初步了解。其中，mapper 文件夹可命名为 dao，两者实际代表的意义相同。pojo 文件夹也可命名为 entity，两者均表示实体类文件夹。

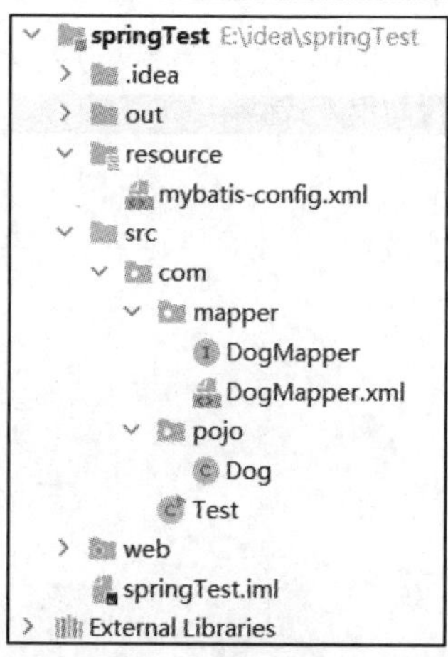

图 7.12 整体目录结构

7.5 本章小结

本章首先介绍了 ORM 框架和 MyBatis 的基本知识，然后介绍了 MyBatis 的工作流程和核心对象，接着通过一个示例讲解了 MyBatis 的简单应用，最后讲解了 MyBatis 的接口开发方式。本章的重点内容是 MyBatis 的简单应用和 MyBatis 的接口开发方式。

在本章内容中，MyBatis 的简单应用是读者应该掌握的部分。另外，在实际开发中，通常使用接口代替 XML 文件进行开发，因此 MyBatis 的接口开发方式也需要读者重点掌握。MyBatis 的工作流程和 SqlSession 对象是面试常考的知识点，读者可按需掌握。

7.6 习题

1. 填空题

（1）MyBatis 通过创建________来创建 SqlSession 对象。

（2）MyBatis 需要配置两种类型的文件，分别是________文件和________文件。

（3）在 MyBatis 映射文件中，负责增删改查的元素分别为________、________、________和________。

2. 选择题

（1）关于 MyBatis，下列描述错误的是（ ）。

A. MyBatis 读取映射文件，获取配置信息

B. MyBatis 根据配置信息创建 SqlSessionFactory 对象

C. SqlSession 对象是提供给用户使用的 API

D. 被处理的 SQL 信息被封装为 SqlSession 对象

（2）关于 MyBatis 接口开发的过程，下列描述错误的是（　　）。

A. MyBatis 的接口名必须和映射文件中的命名空间相同

B. MyBatis 的接口文件和映射文件必须放到同一个目录下

C. 在编写完配置文件与映射文件后，使用 SqlSession 类的 getMapper()方法获取相应的 Mapper 接口的实现类进行操作

D. MyBatis 接口中的方法名必须和映射文件中增删改查操作的 id 值保持一致

3．思考题

（1）简述 MyBatis 的工作流程。

（2）简述 SqlSession 对象在 MyBatis 中的作用。

4．编程题

创建 Dog 类，在其中添加 name 和 age 属性。搭建 MyBatis 框架，通过 MyBatis 框架对 Dog 类进行简单查找操作。

第 8 章 MyBatis 核心配置

本章学习目标

- 了解配置文件与映射文件的结构。
- 掌握 MyBatis 配置文件的使用方法。
- 掌握 MyBatis 映射文件的使用方法。

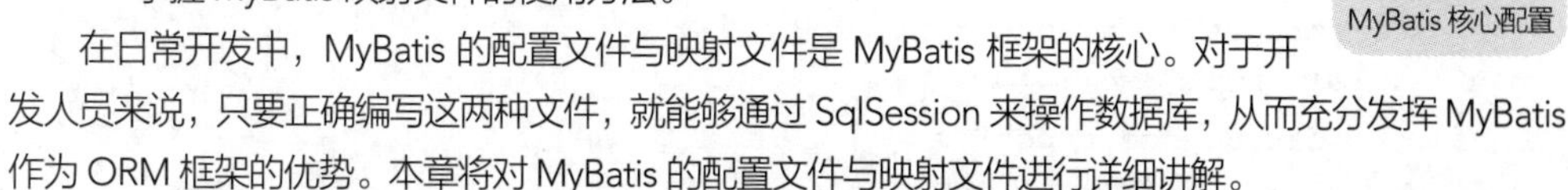

在日常开发中，MyBatis 的配置文件与映射文件是 MyBatis 框架的核心。对于开发人员来说，只要正确编写这两种文件，就能够通过 SqlSession 来操作数据库，从而充分发挥 MyBatis 作为 ORM 框架的优势。本章将对 MyBatis 的配置文件与映射文件进行详细讲解。

8.1 MyBatis 配置文件

MyBatis 配置文件中有 MyBatis 框架的核心配置，负责对 MyBatis 进行全局管理。它包含许多控制 MyBatis 功能的重要元素，读者需要熟练掌握常用元素的使用方法。

8.1.1 配置文件概览

MyBatis 规定了其配置文件的层次结构，如例 8-1 所示。

【例 8-1】 mybatis-config.xml

```
1.  <?xml version="1.0" encoding="UTF-8" ?>
2.  <!DOCTYPE configuration
3.          PUBLIC "-//mybatis.org//DTD Config 3.0//EN"
4.          "http://mybatis.org/dtd/mybatis-3-config.dtd">
5.  <configuration>
6.      <!--设置配置文件-->
7.      <properties>
8.          <property name="" value=""/>
9.      </properties>
10.     <!--MyBatis 设置-->
11.     <settings>
12.         <setting name="" value=""/>
13.     </settings>
14.     <!--包名简化缩写-->
15.     <typeAliases>
16.         <typeAlias type=""/>
17.     </typeAliases>
18.     <!--配置数据类型转换-->
```

```
19.     <typeHandlers>
20.         <typeHandler handler=""/>
21.     </typeHandlers>
22.     <!--自定义结果集对象-->
23.     <objectFactory type=""></objectFactory>
24.     <!--配置插件-->
25.     <plugins>
26.         <plugin interceptor=""></plugin>
27.     </plugins>
28.     <!--配置环境-->
29.     <environments default="">
30.         <!--配置 MySQL 环境-->
31.         <environment id="">
32.             <!--配置事务管理器-->
33.             <transactionManager type=""/>
34.             <!--配置数据库连接-->
35.             <dataSource type="">
36.                 <!--配置数据库连接驱动-->
37.                 <property name="" value=""/>
38.                 <!--配置数据库连接地址-->
39.                 <property name="" value=""/>
40.                 <!--配置用户名-->
41.                 <property name="" value=""/>
42.                 <!--配置密码-->
43.                 <property name="" value=""/>
44.             </dataSource>
45.         </environment>
46.     </environments>
47.     <!--数据厂商标识-->
48.     <databaseIdProvider type=""></databaseIdProvider>
49.     <!--配置 Mapper 映射文件-->
50.     <mappers>
51.         <mapper resource="com/mapper/DogMapper.xml"/>
52.     </mappers>
53. </configuration>
```

例 8-1 中列举了很多关键的元素，这些元素用于支撑 MyBatis 的各项重要功能。需要注意的是，元素在配置文件中的先后顺序是固定的，开发人员需要按照官方提供的顺序编写配置文件，否则会报错。下面针对例 8-1 中的配置文件，详细讲解其中元素的作用和使用方式。

8.1.2 <properties>元素

<properties>元素是一个用于配置属性的元素。MyBatis 支持两种方式配置属性，下面详细介绍这两种方式。

1．在配置文件内设置属性

通过<property>元素，可以将关键的属性添加到配置文件上方，相较于将属性添加到配置文件内部的方式，这种方式更易于维护，具体代码如下所示。

```
<properties>
```

```
    <property name="myDriver" value="com.mysql.cj.jdbc.Driver"/>
    <property name="myUrl" value="jdbc:mysql://127.0.0.1/test?
                        characterEncoding=utf8&amp;
                        useSSL=false&amp;
                        serverTimezone=UTC&amp;
                        allowPublicKeyRetrieval=true"/>
    <property name="myUsername" value="root"/>
    <property name="myPassword" value="root"/>
</properties>
```

在以上代码中，设置了数据库连接的各个属性，在配置数据库连接的相关属性时，可以使用“${}”引用相应属性，引用方式如下所示。

```
<dataSource type="">
    <!--配置数据库连接驱动-->
    <property name="driver" value="${myDriver}"/>
    <!--配置数据库连接地址-->
    <property name="url" value="${myUrl}"/>
    <!--配置用户名-->
    <property name="username" value="${myUsername}"/>
    <!--配置密码-->
    <property name="password" value="${myPassword}"/>
</dataSource>
```

在以上代码中，通过“${}”来引用<property>元素中的 name 属性，引用值为<property>元素中的 value 属性值。

2．配置文件引用外部资源文件

在配置文件中使用<properties>元素的 resource 属性，可以指定需要引入的配置文件。在 src 文件夹下新建 db.properties 文件，如例 8-2 所示。

【例 8-2】 db.properties

```
jdbc.myDriver = com.mysql.cj.jdbc.Driver
jdbc.myUrl = jdbc:mysql://127.0.0.1/test?\
    characterEncoding=utf8&amp;\
    useSSL=false&amp;\
    serverTimezone=UTC&amp;\
    allowPublicKeyRetrieval=true
jdbc.myUsername = root
jdbc.myPassword = root
```

在例 8-2 中，通过 properties 文件设置了数据库连接的相关属性，下面将属性连接到配置文件中，具体如下所示。

```
<properties resource="db.properties"/>
```

配置完成后，可以使用“${}”取用相关属性，具体如下所示。

```
<dataSource type="">
    <!--配置数据库连接驱动-->
    <property name="driver" value="${jdbc.myDriver}"/>
    <!--配置数据库连接地址-->
    <property name="url" value="${jdbc.myUrl}"/>
    <!--配置用户名-->
```

```
    <property name="username" value="${jdbc.myUsername}"/>
    <!--配置密码-->
    <property name="password" value="${jdbc.myPassword}"/>
</dataSource>
```

上述代码使用“${}”引用 db.properties 文件中的内容，完成数据库配置。

8.1.3 <settings>元素

<settings>元素是 MyBatis 中较复杂的元素，这些元素用于控制 MyBatis 运行时的状态和行为。下面讲解<settings>元素的子元素，读者初步了解即可，具体如表 8.1 所示。

表 8.1 <settings>元素的子元素

子元素	说明	有效值
cacheEnabled	配置缓存的全局开关，默认为 true	true、false
lazyLoadingEnabled	延迟加载的全局开关，默认为 false	true、false
aggressiveLazyLoading	当此子元素启用时，任何方法的调用都会触发其所在类中所有属性的加载，默认为 true	true、false
multipleResultSetsEnabled	是否允许单一语句返回多结果集，默认为 true	true、false
useColumnLabel	是否使用列标签代替列名，默认为 true	true、false
useGeneratedKeys	是否使用 JDBC 自动生成主键，默认为 false	true、false
autoMappingBehavior	指定 MyBatis 如何把列映射到字段或属性。NONE 表示取消自动映射；PARTIAL 表示只自动映射没有定义嵌套的结果集，为默认设置；FULL 表示自动映射任意复杂的结果集（无论是否嵌套）	NONE、PARTIAL、FULL
defaultExecutorType	指定执行器类型，默认为 SIMPLE，表示使用普通执行器。除此之外，还可以使用预处理器和批处理器	SIMPLE、REUSE、BATCH
defaultStatementTimeout	设置驱动等待数据库响应的时间，默认没有等待时间	任何正整数
safeRowBoundsEnabled	是否允许在嵌套语句中使用分页，默认为 false	true、false
mapUnderscoreToCamelCase	是否启用自动驼峰命名规则映射，通常数据表的列名采用下画线命名，例如 db_name，需要将其映射为 dbName，此时启用此子元素后，MyBatis 可自动映射，无须手动配置，默认为 false	true、false
localCacheScope	MyBatis 利用本地缓存机制防止循环引用和加速重复嵌套查询。其默认为 SESSION，这种情况下会缓存一个会话中执行的所有查询。若设置其值为 STATEMENT，本地会话仅在语句执行上，对相同 SqlSession 的不同调用将不会共享数据	SESSION、STATEMENT
jdbcTypeForNull	当没有为参数提供特定的 JDBC 类型时，为空值指定 JDBC 类型。某些驱动需要指定列的 JDBC 类型，在多数情况下直接用一般类型即可，默认为 OTHER	OTHER、VARCHAR、NULL
lazyLoadTriggerMethods	指定哪个对象的方法会触发一次延迟加载，默认为 equals、clone、hashCode 和 toString 组成的集合	可选值为一个方法名列表，使用逗号分隔
defaultScriptingLanguage	指定动态 SQL 语句生成的默认语言	可选值为全限定类名
callSettersOnNulls	设置当返回的类型为空时，是否显示空值，默认为 false	true、false
proxyFactory	指定 MyBatis 创建具有延迟加载能力的对象用到的代理工具，默认为 JDK 动态代理	CGLIB、JAVASSIST
logPrefix	指定 MyBatis 增加到日志名称的前缀	任何字符串
logImpl	指定 MyBatis 所用日志的具体实现，通常当需要开启 SQL 日志打印时配置此子元素，默认为 SLF4J	LOG4J、LOG4J2、COMMONS_LOGGING 等

表 8.1 中列举了<settings>元素的子元素，下面给出一个<settings>元素的配置示例，具体代码如下所示。

```
<settings>
   <!--配置缓存的全局开关-->
   <setting name="cacheEnabled" value="true" />
   <!--延迟加载的全局开关-->
   <setting name="lazyLoadingEnabled" value="false" />
   <!--允许单一语句返回多结果集-->
   <setting name="multipleResultSetsEnabled" value="true" />
   <!--使用列标签代替列名，需要兼容驱动-->
   <setting name="useColumnLabel" value="true" />
   <!--允许 JDBC 自动生成主键，需要兼容驱动。如果将此子元素的 value 设置为 true，
   则强制自动生成主键，尽管一些驱动不能兼容，但仍能正常工作-->
   <setting name="useGeneratedKeys" value="false" />
   <!--指定 MyBatis 如何自动映射列到字段或属性-->
   <setting name="autoMappingBehavior" value="PARTIAL" />
   <!—指定执行器类型，此处使用 SIMPLE，即普通执行器-->
   <setting name="defaultExecutorType" value="SIMPLE" />
   <!--设置超时时间，此处设置为 25 秒，25 秒后若没有获得数据库的数据，则终止获取-->
   <setting name="defaultStatementTimeout" value="25"/>
   <!--设置数据库驱动程序默认返回的数据条数-->
   <setting name="defaultFetchSize" value="100" />
   <!--允许在嵌套语句中使用分页（RowBounds）-->
   <setting name="safeRowBoundsEnabled" value="false" />
   <!--启用自动驼峰命名规则映射，即从 a_example 到 aExample 的映射-->
   <setting name="mapUnderscoreToCamelCase" value="true" />
   <!--利用本地缓存机制防止循环引用和加速重复嵌套循环-->
   <setting name="localCacheScope" value="SESSION" />
   <!--指定未设置的 JDBC 类型为 NULL-->
   <setting name="jdbcTypeForNull" value="NULL" />
   <!--指定触发延迟加载的方法-->
   <setting name="lazyLoadTriggerMethods" value="equals" />
</settings>
```

以上是配置<settings>元素的示例，常用的是 mapUnderscoreToCamelCase 子元素的配置，此配置可以自动将数据表的列名转化为以驼峰命名规则命名的名称，从而减少代码量，提高开发效率。表 8.1 中<settings>的子元素读者可以按需学习，不要求全部掌握。

8.1.4 <typeAliases>元素

在开发时，经常需要设置返回结果的类型，以便 MyBatis 进行映射，代码如下所示。

```
<select id="selectByName" resultType="com.Dog">
      select * from dog where name like '%${name}%'
</select>
```

以上代码中的加粗代码用于设置返回值类型，resultType 的值为映射对象相对于 src 文件夹的路径。若此路径过长，会为开发带来不必要的麻烦。为解决此问题，MyBatis 提供了包名缩写的配置，即配置文件中的<typeAliases>元素。<typeAliases>元素允许使用两种方式实现包名缩写。下面详细讲解

<typeAliases>元素的配置方式。

1．<typeAliase>元素

设置别名可简化开发中的配置。使用<typeAliases>元素中的<typeAliase>元素，可为目标类设置别名，代码如下所示。

```
<typeAliases>
    <typeAlias alias="dog" type="com.Dog"/>
</typeAliases>
```

当 mapper 文件中的返回值类型为 Dog 类时，将不再需要书写类名的全路径，mapper 文件中的代码如下所示。

```
<select id="selectByName" resultType="dog">
    select * from dog where name like '%${name}%'
</select>
```

使用<typeAlias>元素可以将 alias 属性的值作为别名，替换 type 属性值中的路径。在实际开发中，实体类非常多，使用这种方式时需要配置大量<typeAlias>元素，这时候就需要使用批量定义别名的方法。

2．<package>元素

<package>元素可以实现批量定义别名的功能，将<package>元素中 name 的值设置为实体类的包名，代码如下所示。

```
<typeAliases>
   <package name="com"/>
</typeAliases>
```

通过这种方式，处于 com 文件夹下的所有类都被配置了别名，别名默认为首字母小写的类名。此方式为开发中常用的别名配置方式，读者需要熟练掌握。

8.1.5 <typeHandlers>元素

在程序运行过程中，当 MyBatis 为 SQL 语句设置参数或者从结果集中取值时，都要通过类型处理器完成数据类型转换。类型处理器是一个负责处理数据库表到实体类之间的转换工具，MyBatis 内部定义了一系列的类型处理器，常用的类型处理器如表 8.2 所示。

表 8.2 常用的类型处理器

类型处理器	Java 类型	JDBC 类型
BooleanTypeHandler	java.lang.Boolean，boolean	boolean
ByteTypeHandler	java.lang.Byte，byte	byte
ShortTypeHandler	java.lang.Short，short	short 或 integer
IntegerTypeHandler	java.lang.Integer，integer	integer
LongTypeHandler	java.lang.Long，long	long 或 integer
FloatTypeHandler	java.lang.Float，float	float
DoubleTypeHandler	java.lang.Double，double	double

表 8.2 中列举了 MyBatis 内部定义的类型处理器，这些类型处理器不需要显示声明，MyBatis 会自动检测数据类型并完成转换。

通常情况下，默认的类型处理器就可以完成大多数场景中数据类型转换的需要。但是，当 JDBC

类型与 Java 类型不匹配时，就需要自定义类型处理器。

自定义类型处理器需要实现 TypeHandler 接口或者继承 BaseTypeHandler 类，编写完相应代码后，通过<typeHandlers>元素将类型处理器配置到 MyBatis，代码如下所示。

```
<typeHandlers>
   <typeHandler
         javaType="Boolean"
         jdbcType="Integer"
         handler="com.BooleanTinyIntTypeHandler"/>
</typeHandlers>
```

在以上代码中，<typeHandler>元素通过 handler 属性指定一个类型处理器。在<typeHandler>元素中，javaType 属性表示转换成 Java 数据时，接收数据的类型；jdbcType 属性表示转换成 JDBC 数据时，接收数据的类型。

8.1.6 <objectFactory>元素

MyBatis 通过 ObjectFactory（对象工厂）创建结果集对象。在默认情况下，MyBatis 通过其定义的 DefaultObjectFactory 类完成相关工作，但是在实际开发中，当需要干预结果集对象的创建过程时，就需要自定义 ObjectFactory。

自定义 ObjectFactory 需要完成两个步骤，首先编写 ObjectFactory 类，使其继承 DefaultObjectFactory 类，然后将 ObjectFactory 类配置到 MyBatis 配置文件中，具体代码如下所示。

```
<objectFactory type="com.DogObjectFactory"/>
```

在以上代码中，通过<objectFactory>元素指定 ObjectFactory，其中，type 属性为 ObjectFactory 类的全限定类名。

8.1.7 <environments>元素

<environments>元素主要的作用是配置数据库连接环境，MyBatis 支持多种环境，在不同的环境中可以操作不同的数据库。通过修改运行环境，MyBatis 可以实现在开发、测试和生产环境中切换数据库连接。

在 MyBatis 配置文件中，利用<environments>元素来配置事务管理器和数据源，具体代码如下所示。

```
<environments default="dev">
   <!--配置 MySQL 环境-->
   <environment id="dev">
      <!--配置事务管理器-->
      <transactionManager type="JDBC"/>
      <!--配置数据库连接-->
      <dataSource type="POOLED">
         <!--配置数据库连接驱动-->
         <property name="driver" value="${myDriver}"/>
         <!--配置数据库连接地址-->
         <property name="url" value="${myUrl}"/>
         <!--配置用户名-->
         <property name="username" value="${myUsername}"/>
         <!--配置密码-->
         <property name="password" value="${myPassword}"/>
```

```
        </dataSource>
    </environment>
</environments>
```

在以上代码中，<environments>元素用于配置环境，在其中可以加入许多<environment>元素，每个<environment>元素都可以设置一个运行环境，在<environments>元素中，可以通过 default 属性来指定默认的运行环境，default 属性的值需要设置为<environment>元素中 id 属性的值。

在<environment>元素中，可以设置两种不同的元素，一种是可作为事务管理器的<transactionManager>元素，另一种是可用于数据库连接的<dataSource>元素。

<transationManager>元素负责为 SQL Map 配置事务管理服务，其中，属性 type 用于指定使用的事务管理器类型，这个属性值可以是一个类名，也可以是一个别名；包含在框架中的 3 个事务管理器分别为 JDBC、JTzA 和 EXTERNAL，日常开发中经常使用 JDBC 事务管理器。

<dataSource>元素负责为 SQL Map 配置数据库连接服务，其中，type 属性用于设置 MyBatis 获取连接的方式，其可选值有 3 个，分别为 UNPOOLED、POOLED 和 JNDI。下面对这 3 个值分别进行讲解。

1．UNPOLLED

UNPOLLED 表示不使用连接池进行连接，即每一次连接时都会新建连接。此方式没有复用资源，但数据源安全可得到保障。

UNPOLLED 类型的数据源需要配置 5 个属性，如表 8.3 所示。

表 8.3　UNPOLLED 数据源需要配置的属性

属性	说明
driver	JDBC 驱动的全限定类名
url	数据库连接地址
username	数据库连接用户名
password	数据库连接密码
defaultTransactionIsolationLevel	默认的连接事务隔离级别

2．POOLED

POOLED 表示使用连接池进行连接，使用此连接方式，每次都需要根据连接池的状态来获取连接，此过程可以分为 3 种情况。

（1）判断空闲连接池内有没有空闲连接，如果有，则返回一个空闲连接。

（2）如果没有空闲连接，则查看活动连接池内是否有空余位置，如果有，则新建一个连接返回，并将此连接加入连接池。

（3）如果活动连接池没有空余位置，则返回在活动连接池内使用最久的连接。

通过此方式配置的数据源相对于通过 UNPOLLED 配置的数据源可以配置更多的属性，POOLED 数据源的常用属性如表 8.4 所示。

表 8.4　POOLED 数据源的常用属性

属性	说明
poolMaximumActiveConnections	可以存在的最大活动连接数量，默认为 10
poolMaximumIdleConnections	可能存在的最大空闲连接数量
poolMaximumCheckoutTime	连接池中的连接被检出的最长时间，默认为 20 秒
poolTimeToWait	重新获取连接需等待的时间，如果获取连接的时间大于这个时间，连接池会打印此次日志，并重新尝试获取一个连接

3．JNDI

JNDI 数据源可连接外部的应用服务器，无须配置连接参数，只需指定外部容器的引用，JNDI 数据源的常用属性只有两个，如表 8.5 所示。

表 8.5　JNDI 数据源的常用属性

属性	说明
initial_context	用来在 InitialContext 中寻找上下文，为可选属性
data_source	用来指定数据源实例位置的上下文路径

此类数据源在开发中不常用，该部分内容读者仅作了解即可。

8.1.8 <mappers>元素

在 MyBatis 配置文件中，<mappers>元素用来引入映射文件。映射文件中包含 POJO 对象和数据表之间的映射信息。<mappers>元素引导 MyBatis 找到并解析相应的配置文件。

<mappers>元素引入映射文件的方法有两种，一种为通过资源路径直接引用，另一种为扫描包中的全部映射器接口进行引用，具体代码如下所示。

```
<mappers>
    <!--引入指定路径的文件-->
    <mapper resource="com/mapper/DogMapper.xml"/>
    <!--将 com.mapper 包中的映射器接口引入-->
    <package name="com.mapper"/>
</mappers>
```

在以上代码中，利用<mapper>元素的 resource 属性指定资源路径并直接引用该路径下的文件；利用<package>元素的 name 属性指定需要扫描的映射器接口所在的包。这两种方式分别为 XML 开发方式与接口开发方式，不可同时使用。

8.2 MyBatis 映射文件

8.2.1 映射文件概述

映射文件是 MyBatis 的重要组成部分，它包含开发中编写的 SQL 语句、参数、结果集等。映射文件需要通过 MyBatis 配置文件中的<mapper>元素引入才能生效。MyBatis 规定了映射文件的层次结构，具体如例 8-3 所示。

【例 8-3】 mapper.xml

```
1.  <?xml version="1.0" encoding="UTF-8"?>
2.  <!DOCTYPE mapper PUBLIC "-//mybatis.org//DTD Mapper 3.0//EN"
3.         "http://mybatis.org/dtd/mybatis-3-mapper.dtd">
4.  <mapper namespace="MyDog">
5.      <!--开启此映射文件的缓存-->
6.      <cache/>
7.      <!--指定引用的命名空间，当此命名空间执行 DML 操作时，
8.      被引用的命名空间中的缓存也会失效-->
```

```
9.      <cache-ref namespace=""/>
10.     <!--参数映射集-->
11.     <parameterMap id="" type="">
12.         <parameter property="" jdbcType="" javaType="" typeHandler=""/>
13.     </parameterMap>
14.     <!--SQL 块-->
15.     <sql id="">
16.     </sql>
17.     <!--映射结果集-->
18.     <resultMap id="" type="">
19.         <id property="" column=""/>
20.         <result property="" column="" />
21.     </resultMap>
22.     <!--查询元素-->
23.     <select id="" resultType="" parameterType="">
24.     </select>
25.    <!--新增元素-->
26.     <insert id="" parameterType="">
27.     </insert>
28.     <!--删除元素-->
29.     <delete id="">
30.     </delete>
31.     <!--更新元素-->
32.     <update id="" parameterType="">
33.     </update>
34.  </mapper>
```

例 8-3 中列举了 MyBatis 映射文件中很多关键的元素，元素在映射文件中的先后顺序是不固定的，开发人员无须按照例 8-3 中的元素顺序编写映射文件。下面详细讲解例 8-3 中的元素及使用方式。

8.2.2 查找元素

<select>元素为查找操作所用的元素，此元素包含查询所用的 SQL 语句、参数类型和返回值类型等信息，具体代码如下所示。

```
 <select id="selectDog" parameterType="dog" resultType="dog">
   select * from dog where name = #{name}
</select>
```

在以上代码中，<select>元素通过 id 属性指定查询的名称，通过 parameterType 属性指定参数类型，通过 resultType 属性指定返回值类型。除此之外，<select>元素还提供了一系列属性，其常用属性如表 8.6 所示。

表 8.6　<select>元素的常用属性

属性	说明
id	用于指定 SQL 语句在命名空间中唯一的标识符
parameterType	用于指定 SQL 语句所需类的完全限定名或别名，此属性在开发中通常不使用
resultType	用于指定返回结果的映射类型
resultMap	用于指定返回结果集的映射，通常在列名与属性不匹配或进行子查询时使用
flushCache	用于设置 SQL 语句被进行之后，是否清空 MyBatis 的本地缓存和二级缓存，默认为 false
useCache	用于设置是否开启 MyBatis 二级缓存，默认为 true，即开启二级缓存

续表

属性	说明
timeout	用于设定驱动程序等待数据库返回请求结果的时间（单位为秒）
fetchSize	用于设置每次可以传输的记录条数
statementType	用于设置 statement 的类型，可选值有 STATEMENT、PREPARED 和 CALLABLE
resultSetType	用于设置 resultSet 的类型
resultOrdered	用于设置在嵌套查询时引用的方式
resultSets	负责列出 SQL 语句执行后返回的每个结果集的名称

表 8.6 中列出了<select>元素的常用属性，读者可根据具体需求选择使用。

8.2.3 增加、删除、修改元素

在 MyBatis 映射文件中，对应增加、删除、修改操作的元素分别是<insert>、<delete>和<update>元素，下面详细讲解这 3 个元素的使用方法。

1．<insert>元素的使用方法

<insert>元素的使用示例如下所示。

```
<insert id="insertDog" parameterType="dog">
  insert into dog(name,age) values(#{name},#{age})
</insert>
```

在以上代码中，<insert>元素通过 parameterType 属性指定参数类型，然后在此元素内部添加 SQL 语句，以上代码执行后，会向数据库中新增一条数据。

2．<delete>元素的使用方法

<delete>元素的使用示例如下所示。

```
<delete id="deleteDog">
  delete from dog where id = #{id}
</delete>
```

在以上代码中，<delete>元素只指定了 id 属性，省略了 parameterType 属性，当传入的参数只有一个时，在<delete>元素内部可以通过“#{}”进行任意赋值。以上代码执行后，会将数据库中对应 id 的数据删除。

3．<update>元素的使用方法

<update>元素的使用示例如下所示。

```
<update id="updateDog" parameterType="dog">
  update dog set name = #{name},age = #{age} where id = #{id}
</update>
```

在以上代码中，<update>元素通过 parameterType 属性指定参数类型，然后在此元素内部添加 SQL 语句，以上代码执行后，会更改数据库中相应 id 的数据。

<insert>、<delete>和<update>元素的返回值均为被修改的数据行数，因此，一般不用设置返回值类型。

8.2.4 结果集元素

<resultMap>元素为映射结果集元素，负责数据库返回值与对象之间的映射，需要读者重点掌握。

在此初步介绍<resultMap>元素，其详细使用方法将在第 9 章讲解。

<resultMap>元素包含一些子元素，这些元素的层次结构如下所示。

```
<resultMap id="" type="">
    <!--设置主键的映射-->
    <id column="" property=""/>
    <!--设置列的映射-->
    <result property="" column=""/>
    <!--设置对象类型映射-->
    <association property="" javaType="">
        <id property="" column=""/>
        <result property="" column=""/>
    </association>
    <!--设置集合类型映射-->
    <collection property="" ofType="">
        <id property="" column=""/>
        <result property="" column=""/>
    </collection>
</resultMap>
```

从以上代码可以看出，<resultMap>元素可进行从数据库数据到对象的映射结果集的设置，其中，id 属性为结果集的名称，type 属性为结果集最终映射的对象。

<id>和<result>元素分别用于设置主键和列的映射，其中，column 属性代表数据表的列名，property 属性代表对象的属性。

<association>和<collection>元素用于进行复杂类型的映射。<association>元素负责列与对象之间的映射，其中 property 属性代表对象的属性，javaType 是类的全限定名。<collection>元素负责列与对象集合的映射，它与<association>元素的不同点为<collection>元素可以映射对象集合，而<association>元素只能映射单一的对象。

8.2.5 <sql>元素

<sql>元素用于简化列名的书写，在同一个命名空间内，重复的列名会给维护带来很多的麻烦。因此，可使用<sql>元素来包含这些列名，使其得到重用，具体代码如下所示。

```
  <sql id="dogCols">
     id,name,age
  </sql>
  <select id="selectDog" parameterType="dog" resultType="dog" >
  select <include refid="dogCols"/> from dog where name = #{name}
</select>
```

以上代码通过<sql>元素来包裹常用的列名，当需要使用此列名时，可以利用<include>元素的 refid 属性来指定。可通过修改<sql>元素实现全文件中此列名的修改。

8.3 本章小结

本章分别对 MyBatis 的配置文件和映射文件进行了讲解，其中，所有元素的使用方法均要求读者掌握。学习完此章，读者可以了解在 MyBatis 框架中类的增删改查方法。

8.4 习题

1．填空题

（1）在 MyBatis 的配置文件中，<typeAlias>元素用于为类设置________。

（2）在 MyBatis 的配置文件中，<environments>元素用于指定________。

（3）在 MyBatis 的映射文件中，用于查询操作的元素是________。

（4）在 MyBatis 的映射文件中，用于定义可重用 SQL 代码片段的元素是________。

2．选择题

（1）关于 MyBatis 的配置文件，下列描述错误的是（　　）。

A. <mapper>元素负责配置 MyBatis 的映射文件

B. <environments>元素负责配置 MyBatis 的数据库连接环境

C. <properties>元素负责配置 MyBatis 的常用属性

D. <settings>元素负责控制 MyBatis 的核心状态和行为

（2）关于 MyBatis 的映射文件，下列描述错误的是（　　）。

A. 一个 MyBatis 配置文件中可以引入多个映射文件

B. 在编写 MyBatis 的配置文件时，开发人员无须关心元素的顺序

C. <mapper>元素是 MyBatis 映射文件的根元素

D. 在编写 MyBatis 映射文件时，可以省略<setting>元素的配置

3．编程题

创建 Dog 类，在其中添加 name 和 age 属性。搭建 MyBatis 框架，通过 MyBatis 框架对 Dog 类进行增删改查操作。

第9章 MyBatis 进阶

本章学习目标

- 了解 MyBatis 缓存的作用。
- 掌握动态 SQL 的使用方法。
- 掌握 MyBatis 的关联映射。
- 掌握 MyBatis 的注解开发。

第 7 章和第 8 章讲解了 MyBatis 的基础知识。本章将对 MyBatis 的进阶知识进行讲解，其中包括 MyBatis 缓存、动态 SQL、MyBatis 关联映射和 MyBatis 注解开发。通过对 MyBatis 进阶知识的了解，读者可以更加熟练地使用 MyBatis。

9.1 MyBatis 缓存

为了减小重复查询给数据库带来的压力，MyBatis 提供了缓存机制，这种机制能够缓存查询的结果，避免重复的查询。通过对本节内容的学习，读者可以了解 MyBatis 的缓存机制，掌握 MyBatis 二级缓存的开启方法。

9.1.1 MyBatis 缓存简介

MyBatis 提供了两种缓存方式：一种为针对 SqlSession 的缓存，此缓存方式默认开启；另一种为针对全局的缓存，需要手动开启。MyBatis 缓存结构如图 9.1 所示。

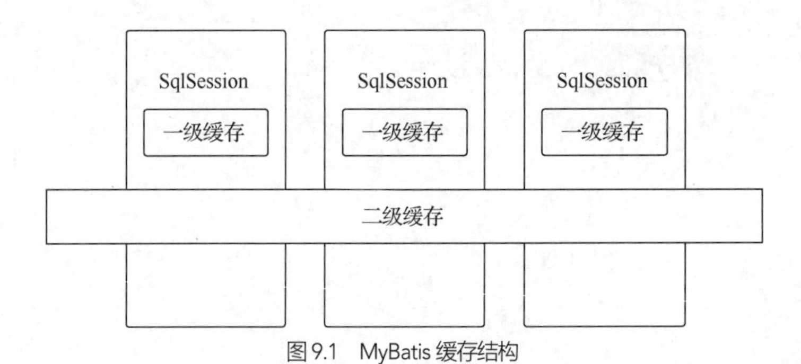

图 9.1 MyBatis 缓存结构

从图 9.1 中可以看到，一级缓存存在 SqlSession 对象中，二级缓存横跨全部的 SqlSession 对象，对所有的查询结果都生效。

9.1.2 一级缓存概述

在没有配置的情况下，MyBatis 默认开启一级缓存。在实际开发时，使用同一个 SqlSession 对象调用同一个 Mapper 方法，往往只执行一次 SQL 语句。这是因为当开启一级缓存时，第一次查询后 MyBatis 会将查询结果放在缓存中，当再次使用这个 SqlSession 对象进行相同查询时，如果数据库的数据没有更改，则直接将缓存中的数据返回，不会再次发送 SQL 到数据库。一级缓存的结构如图 9.2 所示。

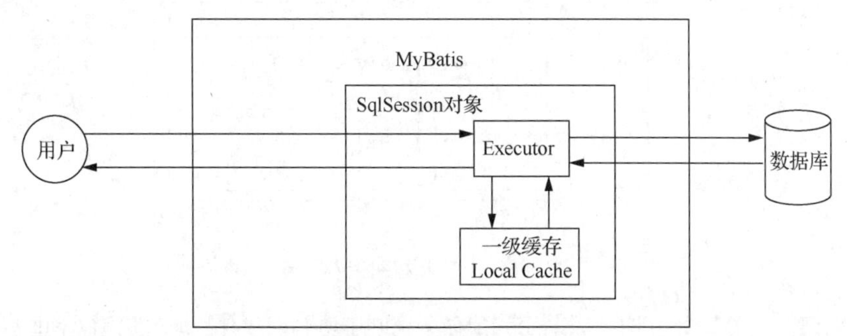

图 9.2 一级缓存的结构

在图 9.2 中，用户发送查询请求给 MyBatis，MyBatis 接收到请求后，创建一个 SqlSession 对象处理本次请求的数据库操作。每个 SqlSession 对象有对应的执行器，执行器在执行 SQL 语句时会查询 Local Cache 中是否存在此查询结果的缓存，如果不存在，则执行此次查询，并将查询结果缓存到 Local Cache 中；如果存在，则直接将此次查询结果的缓存返回。

会话结束时，即调用 SqlSession 对象的 close()方法时，释放此 SqlSession 对象中的所有缓存，并将此 SqlSession 对象禁用。如果要清除缓存中的数据，而不关闭 SqlSession 对象，可以调用 SqlSession 对象的 clearCache()方法，此方法会清空该 SqlSession 对象一级缓存中的所有内容。除此之外，当 SqlSession 对象执行任何一个增加、删除或更改操作时，都将清空此 SqlSession 对象的一级缓存。

在 MyBatis 中，对于两次查询，以下 4 个条件用于判定它们是否是完全相同的查询。

（1）传入的 statementId 是否相同。

（2）查询的结果集范围是否相同。

（3）查询的最终 SQL 语句是否相同。

（4）传递给 Statement 的参数是否相同。

当这些条件的判断结果均为相同时，则认为这两次查询完全相同。

9.1.3 二级缓存概述

MyBatis 的二级缓存是 Application 级别的缓存，与一级缓存的原理类似。不同的是二级缓存的作用域扩大到每个命名空间，在同一个命名空间中的所有查询都将被缓存。二级缓存的结构如图 9.3 所示。

下面详细讲解 MyBatis 二级缓存的执行流程。

（1）MyBatis 中的二级缓存默认关闭，需要手动开启。当开启 MyBatis 二级缓存后，用户发送的有关数据库操作的请求会被 CacheExecutor 拦截，处理后发送给一级缓存，即在二级缓存中查找后，再从一级缓存中查找。

（2）CacheExecutor 拦截数据库操作后，到 Configuration 对象中查看对应命名空间中的缓存，如果发现存在相同查询的缓存，则直接返回该缓存；如果不存在，则进入一级缓存中查找。

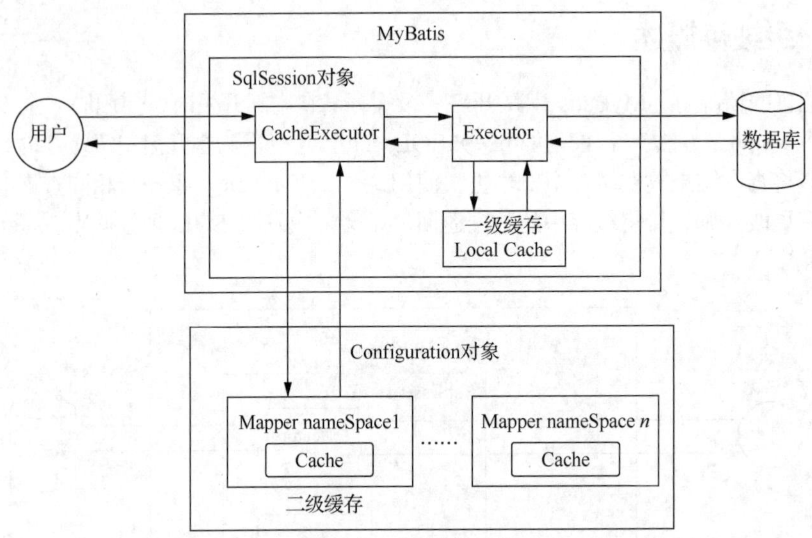

图 9.3　二级缓存的结构

MyBatis 在执行 DML 语句时，会清空当前命名空间中所有的缓存。此外，MyBatis 在开启二级缓存后可能会出现脏读问题：按照开发规范，每个类都有自己的命名空间，在自己的命名空间中不允许有针对其他类的更改，但如果在 B 类的命名空间中对 A 类做出更改，B 类命名空间中的二级缓存会被清除，A 类中的缓存不会被清除；当 A 类命名空间中有针对 A 类的查询操作时，就会寻找二级缓存中是否有相应缓存，如果有，则将其返回。

下面以一个示例讲解二级缓存。

1．创建数据库

准备数据库数据并创建数据库，此例采用例 7-1 中的 dog 表作为实体类，数据如图 9.4 所示。

2．构造目录

构造标准工程目录，具体操作参考 7.4 节。二级缓存目录如图 9.5 所示。

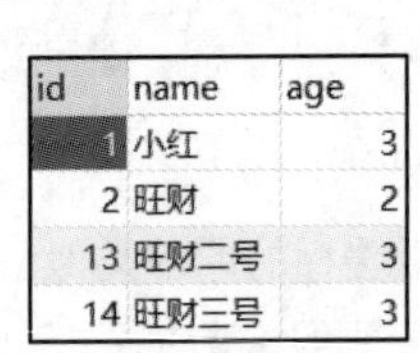

id	name	age
1	小红	3
2	旺财	2
13	旺财二号	3
14	旺财三号	3

图 9.4　dog 表中的数据

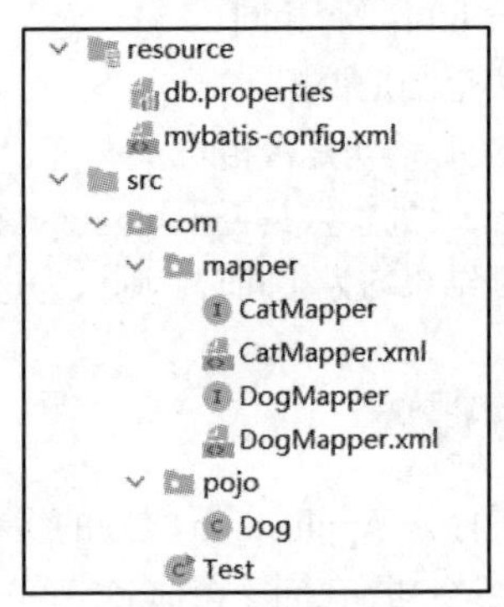

图 9.5　二级缓存目录

3．创建实体类

在 pojo 文件夹下创建 Dog 类，Dog 类如例 9-1 所示（此处省略构造方法、set()方法、get()方法和 toString()方法）。

【例 9-1】 Dog.java

```
1.  public class Dog {
```

```
2.     Integer id;
3.     String name;
4.     Integer age;
5. }
```

此例中的Cat类只用于测试，不创建Cat实体类。

4. 创建配置文件与配置数据库

MyBatis的配置文件如例9-2所示。

【例9-2】 mybatis-config.xml

```
1.  <?xml version="1.0" encoding="UTF-8" ?>
2.  <!DOCTYPE configuration
3.          PUBLIC "-//mybatis.org//DTD Config 3.0//EN"
4.          "http://mybatis.org/dtd/mybatis-3-config.dtd">
5.  <configuration>
6.      <!--引入配置文件-->
7.      <properties resource="db.properties"/>
8.  
9.      <!--设置 -->
10.     <settings>
11.         <!--配置二级缓存的全局开关-->
12.         <setting name="cacheEnabled" value="false" />
13.         <!--开启数据库日志检测-->
14.         <setting name="logImpl" value="STDOUT_LOGGiNG"/>
15.     </settings>
16.     <!--包名简化缩写-->
17.     <typeAliases>
18.         <!--typeAlias 方式-->
19.         <typeAlias alias="dog" type="com.pojo.Dog"/>
20.     </typeAliases>
21.     <!--配置环境-->
22.     <environments default="dev">
23.         <!--配置 MySQL 环境-->
24.         <environment id="dev">
25.             <!--配置事务管理器-->
26.             <transactionManager type="JDBC"/>
27.             <!--配置数据库连接-->
28.             <dataSource type="POOLED">
29.                 <!--配置数据库连接驱动-->
30.                 <property name="driver" value="${jdbc.myDriver}"/>
31.                 <!--配置数据库连接地址-->
32.                 <property name="url" value="${jdbc.myUrl}"/>
33.                 <!--配置用户名-->
34.                 <property name="username" value="${jdbc.myUsername}"/>
35.                 <!--配置密码-->
36.                 <property name="password" value="${jdbc.myPassword}"/>
37.             </dataSource>
38.         </environment>
39.     </environments>
40.     <!--配置 Mapper 映射文件-->
```

```
41.     <mappers>
42.         <!--将 com.mapper 包下的所有 Mapper 接口引入-->
43.         <package name="com.mapper"/>
44.     </mappers>
45. </configuration>
```

在例 9-2 中，首先测试一级缓存的效果，将二级缓存禁用，并开启数据库日志检测。在第 10～15 行代码的<settings>元素中，配置二级缓存的全局开关与 MyBatis 的日志监控。在第 17～20 行代码中使用<typeAliases>元素配置包名缩写，第 43 行代码扫描 Mapper 接口。

在例 9-2 的第 7 行代码中引用的 db.properties 文件如例 9-3 所示。

【例 9-3】 db.properties

```
1.  jdbc.myDriver = com.mysql.cj.jdbc.Driver
2.  jdbc.myUrl = jdbc:mysql://127.0.0.1/test?\
3.    characterEncoding=utf8&\
4.    useSSL=false&\
5.    serverTimezone=UTC&\
6.    allowPublicKeyRetrieval=true
7.  jdbc.myUsername = root
8.  jdbc.myPassword = root
```

5．创建映射文件

在 mapper 文件夹下创建 CatMapper.xml 与 DogMapper.xml 文件，分别如例 9-4、例 9-5 所示。

【例 9-4】 CatMapper.xml

```
1.  <?xml version="1.0" encoding="UTF-8"?>
2.  <!DOCTYPE mapper PUBLIC "-//mybatis.org//DTD Mapper 3.0//EN"
3.          "http://mybatis.org/dtd/mybatis-3-mapper.dtd">
4.  <mapper namespace="com.mapper.CatMapper">
5.      <update id="updateDog">
6.          update dog set age = 40 where id = 1
7.      </update>
8.  </mapper>
```

【例 9-5】 DogMapper.xml

```
1.  <?xml version="1.0" encoding="UTF-8"?>
2.  <!DOCTYPE mapper PUBLIC "-//mybatis.org//DTD Mapper 3.0//EN"
3.          "http://mybatis.org/dtd/mybatis-3-mapper.dtd">
4.  <mapper namespace="com.mapper.DogMapper">
5.      <select id="selectDog" resultType="dog">
6.          select * from dog where id = 1
7.      </select>
8.      <update id="updateDog">
9.          update dog set age = 40 where id = 1
10.     </update>
11. </mapper>
```

在例 9-4 中，使用 CatMapper 的命名空间对 Dog 类进行操作，模拟不规范操作。

6．创建 Mapper 接口文件

在 mapper 文件夹下创建 CatMapper 与 DogMapper 文件，分别如例 9-6 和例 9-7 所示。

【例 9-6】 CatMapper.java

```
1.  public interface CatMapper {
```

```
2.     /*更改*/
3.     Integer updateDog();
4.  }
```

【例 9-7】 DogMapper.java

```
1.  public interface DogMapper {
2.      /*查询*/
3.      Dog selectDog();
4.      /*更改*/
5.      Integer updateDog();
6.  }
```

7. 创建测试类

在 com 文件夹下创建测试类，测试类如例 9-8 所示。

【例 9-8】 Test.java

```
1.  public class Test {
2.      public static void main(String[] args) {
3.          /*创建输入流*/
4.          InputStream inputStream = null;
5.
6.          /*将 MyBatis 配置文件转化为输入流*/
7.          try {
8.              inputStream =
9.                      Resources.getResourceAsStream("mybatis-config.xml");
10.         } catch (IOException e) {
11.             e.printStackTrace();
12.         }
13.
14.         /*通过 SqlSessionFactoryBuilder 类创建 SqlSessionFactory 对象*/
15.         SqlSessionFactory build =
16.                 new SqlSessionFactoryBuilder().build(inputStream);
17.
18.         /*通过 SqlSessionFactory 类创建 SqlSession 对象*/
19.         SqlSession sqlSession = build.openSession();
20.
21.         DogMapper mapper = sqlSession.getMapper(DogMapper.class);
22.
23.         Dog dog1 = mapper.selectDog();
24.         Dog dog2 = mapper.selectDog();
25.
26.         System.out.println(dog1);
27.         System.out.println(dog2);
28.         /*关闭事务*/
29.         sqlSession.close();
30.     }
31. }
```

在例 9-8 中，使用同一个 SqlSession 对象进行多次查询，输出结果如图 9.6 所示。

从图 9.6 中可以看出，查询只进行了一次，此为开启一级缓存的情况。下面使用不同的 SqlSession 对象进行查询，测试一级缓存是否生效，代码如下所示。

```
Checking to see if class com.mapper.CatMapper matches criteria [is assignable to Object]
Checking to see if class com.mapper.DogMapper matches criteria [is assignable to Object]
Opening JDBC Connection
Created connection 380242442.
Setting autocommit to false on JDBC Connection [com.mysql.cj.jdbc.ConnectionImpl@16aa0a0a]
==>  Preparing: select * from dog where id = 1
==> Parameters:
<==    Columns: id, name, age
<==        Row: 1, 小红, 3
<==      Total: 1
Dog{id=1, name='小红', age=3}
Dog{id=1, name='小红', age=3}
Resetting autocommit to true on JDBC Connection [com.mysql.cj.jdbc.ConnectionImpl@16aa0a0a]
Closing JDBC Connection [com.mysql.cj.jdbc.ConnectionImpl@16aa0a0a]
Returned connection 380242442 to pool.

Process finished with exit code 0
```

图9.6　第一次测试结果

```
1.    /*通过SqlSessionFactory类创建SqlSession对象*/
2.     SqlSession sqlSession1 = build.openSession();
3.     SqlSession sqlSession2 = build.openSession();
4.
5.     DogMapper mapper1 = sqlSession1.getMapper(DogMapper.class);
6.     DogMapper mapper2 = sqlSession2.getMapper(DogMapper.class);
7.
8.     Dog dog1 = mapper1.selectDog();
9.     Dog dog2 = mapper2.selectDog();
10.
11.    System.out.println(dog1);
12.    System.out.println(dog2);
13.    /*关闭事务*/
14.    sqlSession1.close();
15.    sqlSession2.close();
16.
```

在以上代码中，使用不同的SqlSession对象进行相同的查询，结果如图9.7所示。

```
Checking to see if class com.mapper.CatMapper matches criteria [is assignable to Object]
Checking to see if class com.mapper.DogMapper matches criteria [is assignable to Object]
Opening JDBC Connection
Created connection 398690014.
Setting autocommit to false on JDBC Connection [com.mysql.cj.jdbc.ConnectionImpl@17c386de]
==>  Preparing: select * from dog where id = 1
==> Parameters:
<==    Columns: id, name, age
<==        Row: 1, 小红, 3
<==      Total: 1
Opening JDBC Connection
Created connection 1543148593.
Setting autocommit to false on JDBC Connection [com.mysql.cj.jdbc.ConnectionImpl@5bfa9431]
==>  Preparing: select * from dog where id = 1
==> Parameters:
<==    Columns: id, name, age
<==        Row: 1, 小红, 3
<==      Total: 1
Dog{id=1, name='小红', age=3}
Dog{id=1, name='小红', age=3}
Resetting autocommit to true on JDBC Connection [com.mysql.cj.jdbc.ConnectionImpl@17c386de]
Closing JDBC Connection [com.mysql.cj.jdbc.ConnectionImpl@17c386de]
Returned connection 398690014 to pool.
Resetting autocommit to true on JDBC Connection [com.mysql.cj.jdbc.ConnectionImpl@5bfa9431]
Closing JDBC Connection [com.mysql.cj.jdbc.ConnectionImpl@5bfa9431]
Returned connection 1543148593 to pool.

Process finished with exit code 0
```

图9.7　第二次测试结果

从图 9.7 中可以看出，查询进行了两次，一级缓存没有起作用。下面在 MyBatis 配置文件中开启二级缓存，代码如下。

```
<!--配置二级缓存的全局开关-->
<setting name="cacheEnabled" value="true" />
```

开启二级缓存后执行测试类，输出结果如图 9.8 所示。

```
Checking to see if class com.mapper.CatMapper matches criteria [is assignable to Object]
Checking to see if class com.mapper.DogMapper matches criteria [is assignable to Object]
Opening JDBC Connection
Created connection 380242442.
Setting autocommit to false on JDBC Connection [com.mysql.cj.jdbc.ConnectionImpl@16aa0a0a]
==>  Preparing: select * from dog where id = 1
==> Parameters: 
<==    Columns: id, name, age
<==        Row: 1, 小红, 3
<==      Total: 1
Dog{id=1, name='小红', age=3}
Dog{id=1, name='小红', age=3}
Resetting autocommit to true on JDBC Connection [com.mysql.cj.jdbc.ConnectionImpl@16aa0a0a]
Closing JDBC Connection [com.mysql.cj.jdbc.ConnectionImpl@16aa0a0a]
Returned connection 380242442 to pool.

Process finished with exit code 0
```

图 9.8　第三次测试结果

从图 9.8 中可以看出，查询只进行了一次，二级缓存生效。下面测试 DML 操作将二级缓存清空的情况，代码如下所示。

```
1.      /*通过 SqlSessionFactory 类创建 SqlSession 对象*/
2.      SqlSession sqlSession1 = build.openSession();
3.      SqlSession sqlSession2 = build.openSession();
4.
5.      /*使用两个不同的 Sqlsession 对象进行操作，一级缓存失效*/
6.      DogMapper dogMapper1 = sqlSession1.getMapper(DogMapper.class);
7.      DogMapper dogMapper2 = sqlSession2.getMapper(DogMapper.class);
8.
9.      /*使用两个不同的 sqlSession 对象在同一个命名空间内进行操作，
10.     其中一个更改了数据库的数据，观察另一个是否使用缓存数据*/
11.     Dog dog1 = dogMapper1.selectDog();
12.
13.     dogMapper2.updateDog();
14.     sqlSession2.commit();
15.
16.     Dog dog2 = dogMapper2.selectDog();
17.
18.     System.out.println(dog1);
19.     System.out.println(dog2);
20.     /*关闭事务*/
21.     sqlSession1.close();
22.     sqlSession2.close();
```

以上代码使用两个不同的 SqlSession 对象进行操作，其中一个 SqlSession 对象更改了数据库的数据，观察另一个 SqlSession 对象在查询相同的数据时是否使用缓存数据，输出结果如图 9.9 所示。

```
Checking to see if class com.mapper.CatMapper matches criteria [is assignable to Object]
Checking to see if class com.mapper.DogMapper matches criteria [is assignable to Object]
Opening JDBC Connection
Created connection 1606286799.
Setting autocommit to false on JDBC Connection [com.mysql.cj.jdbc.ConnectionImpl@5fbdfdcf]
==>  Preparing: select * from dog where id = 1
==> Parameters:
<==    Columns: id, name, age
<==        Row: 1, 小红, 3
<==      Total: 1
Opening JDBC Connection
Created connection 1293680734.
Setting autocommit to false on JDBC Connection [com.mysql.cj.jdbc.ConnectionImpl@4d1c005e]
==>  Preparing: update dog set age = 40 where id = 1
==> Parameters:
<==    Updates: 1
Committing JDBC Connection [com.mysql.cj.jdbc.ConnectionImpl@4d1c005e]
==>  Preparing: select * from dog where id = 1
==> Parameters:
<==    Columns: id, name, age
<==        Row: 1, 小红, 40
<==      Total: 1
Dog{id=1, name='小红', age=3}
Dog{id=1, name='小红', age=40}
Resetting autocommit to true on JDBC Connection [com.mysql.cj.jdbc.ConnectionImpl@5fbdfdcf]
Closing JDBC Connection [com.mysql.cj.jdbc.ConnectionImpl@5fbdfdcf]
Returned connection 1606286799 to pool.
Resetting autocommit to true on JDBC Connection [com.mysql.cj.jdbc.ConnectionImpl@4d1c005e]
Closing JDBC Connection [com.mysql.cj.jdbc.ConnectionImpl@4d1c005e]
Returned connection 1293680734 to pool.

Process finished with exit code 0
```

图 9.9　第四次测试结果

从图 9.9 中可以看出，两次查询的结果不相同。在第二个 SqlSession 对象查询时，二级缓存失效，重新在数据库中查询。下面测试开启 MyBatis 二级缓存后出现的脏读情况，代码如下所示。

```
1.     /*通过 SqlSessionFactory 类创建 SqlSession 对象*/
2.     SqlSession sqlSession1 = build.openSession();
3.     SqlSession sqlSession2 = build.openSession();
4.     /*使用两个不同的 SqlSession 对象进行操作，一级缓存失效*/
5.     DogMapper dogMapper1 = sqlSession1.getMapper(DogMapper.class);
6.     CatMapper catMapper2 = sqlSession2.getMapper(CatMapper.class);
7.     /*使用两个不同的 sqlSession 对象在同一个命名空间内进行操作，
8.     其中一个更改了数据库的数据，观察另一个是否使用缓存数据*/
9.     Dog dog1 = dogMapper1.selectDog();
10.    catMapper2.updateDog();
11.    /*注意，两个 SqlSession 对象不同，因此需要先提交 SqlSession2 的操作*/
12.    sqlSession2.commit();
13.    Dog dog2 = dogMapper1.selectDog();
14.
15.    System.out.println(dog1);
16.    System.out.println(dog2);
17.
18.    /*关闭事务*/
19.    sqlSession1.close();
20.    sqlSession2.close();
```

在以上代码中，使用第二个 SqlSession 对象在 Cat 类命名空间中对 Dog 类进行更改，随后使用第一个 SqlSession 对象进行查询。如果 Cat 类进行操作，清除了 Dog 类的命名空间中的缓存，此时查询的结果数据应该为最新数据。输出结果如图 9.10 所示。

```
Checking to see if class com.mapper.CatMapper matches criteria [is assignable to Object]
Checking to see if class com.mapper.DogMapper matches criteria [is assignable to Object]
Opening JDBC Connection
Created connection 1325144078.
Setting autocommit to false on JDBC Connection [com.mysql.cj.jdbc.ConnectionImpl@4efc180e]
==>  Preparing: select * from dog where id = 1
==> Parameters: 
<==    Columns: id, name, age
<==        Row: 1, 小红, 3
<==      Total: 1
Opening JDBC Connection
Created connection 138817329.
Setting autocommit to false on JDBC Connection [com.mysql.cj.jdbc.ConnectionImpl@8462f31]
==>  Preparing: update dog set age = 40 where id = 1
==> Parameters: 
<==    Updates: 1
Committing JDBC Connection [com.mysql.cj.jdbc.ConnectionImpl@8462f31]
Dog{id=1, name='小红', age=3}
Dog{id=1, name='小红', age=3}
Resetting autocommit to true on JDBC Connection [com.mysql.cj.jdbc.ConnectionImpl@4efc180e]
Closing JDBC Connection [com.mysql.cj.jdbc.ConnectionImpl@4efc180e]
Returned connection 1325144078 to pool.
Resetting autocommit to true on JDBC Connection [com.mysql.cj.jdbc.ConnectionImpl@8462f31]
Closing JDBC Connection [com.mysql.cj.jdbc.ConnectionImpl@8462f31]
Returned connection 138817329 to pool.

Process finished with exit code 0
```

图 9.10　第五次测试结果

从图 9.10 可以看出，查询操作只进行了一次，第二次的查询并没有进行，直接使用了缓存数据，这说明进行更新操作后并没有清除 Dog 类命名空间中的缓存。但在第二次查询时，数据库的数据已经更改，所以此次查询到的数据为脏数据。要想避免此类情况的发生，就需要遵守开发规范，避免跨命名空间的操作。

9.2 动态 SQL

在实际开发中，经常遇到复杂 SQL 构造的情况，开发人员需要使用业务代码对 SQL 语句进行拼接，然后将拼接好的 SQL 语句作为参数传给 MyBatis，这种做法较复杂，且代码不易维护。为解决此类问题，MyBatis 提供了动态 SQL 来灵活拼接 SQL 语句。本节将详细讲解动态 SQL 的使用方法，读者需要熟练掌握本节全部内容。

9.2.1 动态 SQL 简述

动态 SQL 是 MyBatis 提供的拼接 SQL 语句的强大机制。在 MyBatis 的映射文件中，开发人员可通过动态 SQL 灵活组装 SQL 语句，避免业务组装，提高了 SQL 语句的灵活性。

MyBatis 提供了一系列的动态 SQL 元素，其中常见的元素如表 9.1 所示。

表 9.1　常见的动态 SQL 元素

元素	说明
<if>	判断某个条件是否符合，符合则拼接此 SQL 语句
<choose>、<when>、<otherwise>	判断多个条件是否符合，符合则拼接此 SQL 语句
<trim>、<where>、<set>	限定语句，用于限制 SQL 语句的格式
<foreach>	循环语句，用于循环拼接 SQL 语句
<bind>	命名元素，用于创建一个变量，以便后续重复使用

下面详细介绍表 9.1 中的动态 SQL 元素。

9.2.2 <if>元素

<if>元素主要用于条件判断，如果符合此条件，则拼接此元素中的 SQL 语句。<if>元素常用于 where 子句中条件的筛选。例如，有一张 dog 表，其中有 name、age 和其他属性字段，当需要通过 name 和 age 属性查询某数据时，编写的 SQL 语句如下所示。

```
select * from dog where name = #{name} and age = #{age}
```

在以上代码中，如果 name 属性与 age 属性中有一个为空，查询语句就会报错。为解决此问题，需要判断 name 与 age 属性是否为空，如果为空则不进行对应的筛选条件判断。利用<if>元素可以判断属性是否为空。下面以一个示例讲解<if>元素的使用方法，之后元素的相关讲解也将使用这个示例。

1．数据准备

在数据库 test 中创建数据表 dog，SQL 语句如下所示。

```
1.  DROP TABLE IF EXISTS 'dog';
2.  CREATE TABLE 'dog'  (
3.    'id' int(0) NOT NULL AUTO_INCREMENT,
4.    'name' varchar(255) CHARACTER,
5.    'age' int(0) NULL DEFAULT NULL,
6.    PRIMARY KEY ('id') USING BTREE
7.  )
```

向 dog 表中插入数据，SQL 语句如下所示。

```
1.  INSERT INTO 'dog' VALUES (1, '小红', 4);
2.  INSERT INTO 'dog' VALUES (2, '旺财', 2);
3.  INSERT INTO 'dog' VALUES (3, '胖胖', 3);
```

查看 dog 表的结构与数据，如图 9.11 所示。

从图 9.11 中可以看出，表创建完毕，数据插入成功。

2．创建目录结构

在 src 文件夹下创建 com 文件夹，在 com 文件夹下创建 mapper 文件夹和 pojo 文件夹，在项目文件夹下创建 resource 文件夹，并把其标记为资源文件夹，如图 9.12 所示。

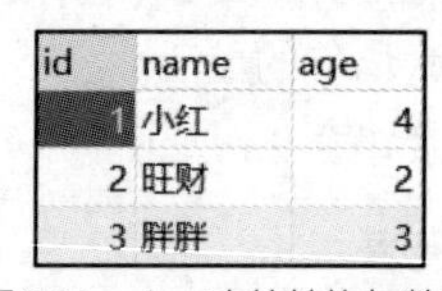

id	name	age
1	小红	4
2	旺财	2
3	胖胖	3

图 9.11　dog 表的结构与数据

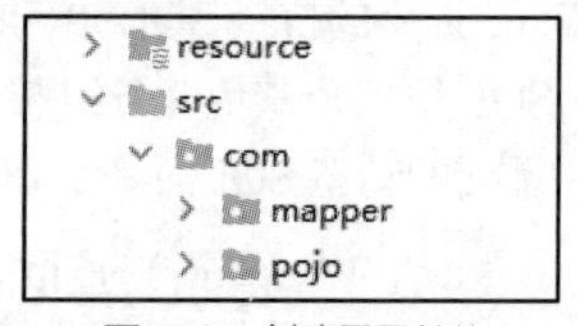

图 9.12　创建目录结构

3．创建实体类

在 pojo 文件夹下创建 Dog 类，代码如例 9-9 所示（此处省略 toString()、get()、set()和构造方法）。

【例 9-9】 Dog.java

```
1.  public class Dog {
2.      Integer id;
```

```
3.      String name;
4.      Integer age;
5.  }
```

4．创建配置文件和数据库配置文件

在 resource 文件夹下创建 MyBatis 配置文件，如例 9-10 所示。

【例 9-10】 mybatis-config.xml

```
1.  <?xml version="1.0" encoding="UTF-8" ?>
2.  <!DOCTYPE configuration
3.          PUBLIC "-//mybatis.org//DTD Config 3.0//EN"
4.          "http://mybatis.org/dtd/mybatis-3-config.dtd">
5.  <configuration>
6.      <!--引入配置文件-->
7.      <properties resource="db.properties"/>
8.  
9.      <!--设置 -->
10.     <settings>
11.         <!--开启数据库日志检测-->
12.         <setting name="logImpl" value="STDOUT_LOGGiNG"/>
13.     </settings>
14.     <!--包名简化缩写-->
15.     <typeAliases>
16.         <!--typeAlias 方式-->
17.         <typeAlias alias="dog" type="com.pojo.Dog"/>
18.     </typeAliases>
19.     <!--配置环境-->
20.     <environments default="dev">
21.         <!--配置 MySQL 环境-->
22.         <environment id="dev">
23.             <!--配置事务管理器-->
24.             <transactionManager type="JDBC"/>
25.             <!--配置数据库连接-->
26.             <dataSource type="POOLED">
27.                 <!--配置数据库连接驱动-->
28.                 <property name="driver" value="${jdbc.myDriver}"/>
29.                 <!--配置数据库连接地址-->
30.                 <property name="url" value="${jdbc.myUrl}"/>
31.                 <!--配置用户名-->
32.                 <property name="username" value="${jdbc.myUsername}"/>
33.                 <!--配置密码-->
34.                 <property name="password" value="${jdbc.myPassword}"/>
35.             </dataSource>
36.         </environment>
37.     </environments>
38.     <!--配置 Mapper 映射文件-->
39.     <mappers>
40.         <!--将 com.mapper 包下的所有 Mapper 接口引入-->
41.         <package name="com.mapper"/>
42.     </mappers>
43. </configuration>
```

例 9-10 中的第 7 行代码引入数据库配置文件，数据库配置文件如例 9-11 所示。

【例 9-11】 db.properties

```
1.  jdbc.myDriver = com.mysql.cj.jdbc.Driver
2.  jdbc.myUrl = jdbc:mysql://127.0.0.1/test?\
3.    characterEncoding=utf8&\
4.    useSSL=false&\
5.    serverTimezone=UTC&\
6.    allowPublicKeyRetrieval=true
7.  jdbc.myUsername = root
8.  jdbc.myPassword = root
```

5．创建映射文件与 Mapper 接口文件

在 mapper 文件夹下创建 DogMapper.xml 文件，其代码如例 9-12 所示。

【例 9-12】 DogMapper.xml

```
1.  <?xml version="1.0" encoding="UTF-8"?>
2.  <!DOCTYPE mapper PUBLIC "-//mybatis.org//DTD Mapper 3.0//EN"
3.          "http://mybatis.org/dtd/mybatis-3-mapper.dtd">
4.  <mapper namespace="com.mapper.DogMapper">
5.      <select id="selectDog" resultType="dog">
6.          select * from dog where
7.          <if test="null != name and '' != name">
8.              name = #{name}
9.          </if>
10.         <if test="null != age and '' != age">
11.             and age = #{age}
12.         </if>
13.     </select>
14. </mapper>
```

在例 9-12 的第 7 行代码中，使用<if>元素控制 SQL 语句是否拼接，其中，test 属性用于设置判断条件，如有多个判断条件，使用 and 连接。

根据 DogMapper.xml 文件编写接口文件，Mapper 接口文件如例 9-13 所示。

【例 9-13】 DogMapper.java

```
1.  package com.mapper;
2.
3.  import com.pojo.Dog;
4.  import java.util.Map;
5.
6.  public interface DogMapper {
7.      /*查询*/
8.      Dog selectDog(Map dog);
9.  }
```

在例 9-13 中，使用 Map 类型的参数，其中需要添加 name 与 age 键值对。

6．编写测试类

在 com 文件夹下创建测试文件，具体代码如例 9-14 所示。

【例 9-14】 Test.java

```
1.  public class Test {
2.      public static void main(String[] args) {
```

```
3.          /*创建输入流*/
4.          InputStream inputStream = null;
5.
6.          /*将 MyBatis 配置文件转化为输入流*/
7.          try {
8.             inputStream =
9.                    Resources.getResourceAsStream("mybatis-config.xml");
10.         } catch (IOException e) {
11.            e.printStackTrace();
12.         }
13.
14.         /*通过 SqlSessionFactoryBuilder 类创建 SqlSessionFactory 对象*/
15.         SqlSessionFactory build =
16.               new SqlSessionFactoryBuilder().build(inputStream);
17.
18.         /*通过 SqlSessionFactory 类创建 SqlSession 对象*/
19.         SqlSession sqlSession = build.openSession();
20.         /*获取 MyBatis 提供的实现类*/
21.         DogMapper mapper = sqlSession.getMapper(DogMapper.class);
22.         /*调用查询方法*/
23.         HashMap hashMap = new HashMap();
24.         hashMap.put("name","小红");
25.         hashMap.put("age",40);
26.
27.         Dog dog1 = mapper.selectDog(hashMap);
28.
29.         System.out.println(dog1);
30.
31.         /*关闭事务*/
32.         sqlSession.close();
33.     }
34. }
```

例 9-14 中的第 27 行代码调用查询方法，传入封装好的参数。执行测试类，输出结果如图 9.13 所示。

```
Checking to see if class com.mapper.DogMapper matches criteria [is assignable to Object]
Opening JDBC Connection
Created connection 1704237553.
Setting autocommit to false on JDBC Connection [com.mysql.cj.jdbc.ConnectionImpl@659499f1]
==>  Preparing: select * from dog where name = ? and age = ?
==> Parameters: 小红(String), 40(Integer)
<==    Columns: id, name, age
<==        Row: 1, 小红, 40
<==      Total: 1
Dog{id=1, name='小红', age=40}
Resetting autocommit to true on JDBC Connection [com.mysql.cj.jdbc.ConnectionImpl@659499f1]
Closing JDBC Connection [com.mysql.cj.jdbc.ConnectionImpl@659499f1]
Returned connection 1704237553 to pool.

Process finished with exit code 0
```

图 9.13　<if>元素测试结果

从图 9.13 中可以看出，name 与 age 拼接，组成完整的 SQL 语句。下面测试 age 属性为空的情况。将例 9-14 测试类中的第 25 行代码删除，使传入的 age 属性为空，执行测试类，输出结果如图 9.14 所示。

```
Checking to see if class com.mapper.DogMapper matches criteria [is assignable to Object]
Opening JDBC Connection
Created connection 1206051975.
Setting autocommit to false on JDBC Connection [com.mysql.cj.jdbc.ConnectionImpl@47e2e487]
==>  Preparing: select * from dog where name = ?
==> Parameters: 小红(String)
<==    Columns: id, name, age
<==        Row: 1, 小红, 40
<==      Total: 1
Dog{id=1, name='小红', age=40}
Resetting autocommit to true on JDBC Connection [com.mysql.cj.jdbc.ConnectionImpl@47e2e487]
Closing JDBC Connection [com.mysql.cj.jdbc.ConnectionImpl@47e2e487]
Returned connection 1206051975 to pool.

Process finished with exit code 0
```

图 9.14　<if>元素空值测试结果

从图 9.14 中可以看出，拼接的 SQL 语句中没有 age 属性。在此需要注意的是，如果 name 属性为空，而 age 属性不为空，或 name 属性和 age 属性都为空，将会报错。拼接出的错误 SQL 语句如下所示。

```
/*当 name 属性为空，age 属性不为空*/
select * from dog where and age = 40
/*当 name 属性、age 属性都为空*/
select * from dog where
```

以上的 SQL 拼接语句不符合标准，此问题可通过<where>元素来解决。

9.2.3　<where>、<set>、<trim>元素

在 9.2.2 小节中，只使用<if>元素不能解决空属性筛选的问题，本小节中将利用<where>元素解决此问题。下面将通过几个示例介绍<where>、<set>、<trim>元素的使用方法。

1. <where>元素

当<where>元素中存在内容时，自动拼接“where”字符串，并且将<where>元素中的第一个“and”字符串删除；当<where>元素中没有内容时，将删除<where>元素自身，即不拼接“where”字符串。修改例 9-12 中映射文件的代码，具体代码如下所示。

```
1.  <select id="selectDog" resultType="dog">
2.      select * from dog
3.      <where>
4.          <if test="null != name and '' != name">
5.              and name = #{name}
6.          </if>
7.          <if test="null != age and '' != age">
8.              and age = #{age}
9.          </if>
10.     </where>
11. </select>
```

在以上代码中，使用<where>元素包裹<if>元素，当两个<if>元素都失效时，得到的 SQL 语句如下所示。

```
select * from dog
```

当 name 属性或 age 属性为空时，得到的 SQL 语句如下所示。

```
/*当 name 属性为空，age 属性不为空*/
select * from dog where age= #{age}
/*当 age 属性为空，name 属性不为空*/
select * from dog where name = #{name}
```

将例 9-14 中测试类的第 24 行代码删除，使传入的 name 属性为空，执行测试类，输出结果如图 9.15 所示。

```
Checking to see if class com.mapper.DogMapper matches criteria [is assignable to Object]
Opening JDBC Connection
Created connection 454104863.
Setting autocommit to false on JDBC Connection [com.mysql.cj.jdbc.ConnectionImpl@1b11171f]
==>  Preparing: select * from dog WHERE age = ?
==> Parameters: 40(Integer)
<==    Columns: id, name, age
<==        Row: 1, 小红, 40
<==      Total: 1
Dog{id=1, name='小红', age=40}
Resetting autocommit to true on JDBC Connection [com.mysql.cj.jdbc.ConnectionImpl@1b11171f]
Closing JDBC Connection [com.mysql.cj.jdbc.ConnectionImpl@1b11171f]
Returned connection 454104863 to pool.
```

图 9.15 <where>元素测试结果

由图 9.15 可知，空属性筛选问题得到解决。

2. <set>元素

<set>元素用于更新操作的 SQL 语句的拼接。当<set>元素中存在内容时，自动拼接“set”字符串，并且将<set>元素中最后的“,”字符串删除；当<set>元素中没有内容时，将删除<set>元素自身，即不拼接“set”字符串。此处需要注意的是，当“set”字符串被删除时，会报错，此时可以在<set>元素中添加“id=#{id}”使 SQL 语句始终有效。修改例 9-12 映射文件中的代码，调整后的代码如下所示。

```
1.  <update id="updateDog" >
2.        update dog
3.        <set>
4.            <if test="null != name and '' != name">
5.                name = #{name},
6.            </if>
7.            <if test="null != age and '' != age">
8.                age = #{age},
9.            </if>
10.       </set>
11.       where id = #{id}
12. </update>
```

在以上代码中，使用<set>元素包裹<if>元素，无论哪个<if>元素失效，都可以正常拼接 SQL 语句。以将 name 属性设为空值、为 age 属性赋值为例，修改 Mapper 接口的代码，在例 9-13 的代码中添加更新方法，调整后的代码如下所示。

```
/*更新*/
Integer updateDog(Map dog);
```

修改测试类代码，代码如下所示。

```
1.  /*调用更新方法*/
2.  HashMap hashMap = new HashMap();
3.  hashMap.put("age",3);
```

```
4.   hashMap.put("id",1);
5.
6.   mapper.updateDog(hashMap);
```

在以上代码中，不为 name 属性赋值，修改 id 属性的值为 1，将 age 属性设置为 3。执行测试类，输出结果如图 9.16 所示。

```
Checking to see if class com.mapper.DogMapper matches criteria [is assignable to Object]
Opening JDBC Connection
Created connection 538592647.
Setting autocommit to false on JDBC Connection [com.mysql.cj.jdbc.ConnectionImpl@201a4587]
==>  Preparing: update dog SET age = ? where id = ?
==> Parameters: 3(Integer), 1(Integer)
<==    Updates: 1
Rolling back JDBC Connection [com.mysql.cj.jdbc.ConnectionImpl@201a4587]
Resetting autocommit to true on JDBC Connection [com.mysql.cj.jdbc.ConnectionImpl@201a4587]
Closing JDBC Connection [com.mysql.cj.jdbc.ConnectionImpl@201a4587]
Returned connection 538592647 to pool.

Process finished with exit code 0
```

图 9.16　<set>元素测试结果

从图 9.16 中可以看出，SQL 语句拼接正常。

3．<trim>元素

<trim>元素的使用方法较复杂，拼装 SQL 语句的灵活性极强，可以模拟<where>元素与<set>元素的功能。<trim>元素提供了许多属性来控制 SQL 语句的拼装，常用属性如表 9.2 所示。

表 9.2　<trim>元素的常用属性

属性	说明
prefix	在拼接内部元素时，在需要拼接的 SQL 语句前面额外附加字符串
prefixOverride	在拼接内部元素时，在需要拼接的 SQL 语句前面删除字符串
suffix	在拼接内部元素时，在需要拼接的 SQL 语句后面额外附加字符串
suffixOverride	在拼接内部元素时，在需要拼接的 SQL 语句后面删除字符串

使用<trim>元素模拟<set>元素的功能，在例 9-12 的代码中添加<update>元素，需要添加的代码如下所示。

```
1.   <update id="updateDog" >
2.       update dog
3.       <trim prefix="set" suffixOverrides=",">
4.           <if test="null != name and '' != name">
5.               name = #{name},
6.           </if>
7.           <if test="null != age and '' != age">
8.               age = #{age},
9.           </if>
10.      </trim>
11.      where id = #{id}
12.  </update>
```

利用<trim>元素的 prefix 属性，为需要拼接的 SQL 语句添加“set”字符串前缀；利用<trim>元素的 suffixOverrides 属性，将 SQL 语句最后的“,”字符串删除，完成<set>元素功能的模拟。

下面使用<trim>元素模拟<where>元素的功能，修改例 9-12 中的代码，调整后的代码如下所示。

```
1.   <select id="selectDogTrim" resultType="dog">
```

```
2.      select * from dog
3.      <trim prefixOverrides="and" prefix="where">
4.         <if test="null != name and '' != name">
5.            and name = #{name}
6.         </if>
7.         <if test="null != age and '' != age">
8.            and age = #{age}
9.         </if>
10.     </trim>
11. </select>
```

利用<trim>元素的 prefix 属性，为需要拼接的 SQL 语句添加“where”字符串前缀；利用<trim>元素的 prefixOverrides 属性，将 SQL 语句前面的“and”字符串删除，完成<where>元素功能的模拟。

9.2.4 <choose>、<when>、<otherwise>元素

在开发中经常遇到当某一个条件不成立时执行另一个操作的情况，这时就要使用<choose>、<when>和<otherwise>元素。修改例 9-12 中的代码，调整后的代码如下所示。

```
1.  <select id="selectDogChoose" resultType="dog">
2.      select * from dog
3.      <where>
4.         <choose>
5.            <when test="null != name">
6.               and name = #{name}
7.            </when>
8.            <when test="null != age and '' != age">
9.               and age = #{age}
10.           </when>
11.           <otherwise>
12.              and id = #{id}
13.           </otherwise>
14.        </choose>
15.     </where>
16. </select>
```

<choose>元素可以选择拼接其中一段代码，其中需要使用<when>元素判断是否拼接此段代码，从上至下，只要有一个<when>元素符合条件，便拼接此段代码，然后退出<choose>元；当所有<when>元素都不符合条件时，拼接<otherwise>元素中的代码。

9.2.5 <foreach>元素

<foreach>元素可以循环拼接 SQL 语句，例如，拼接查询 id 为 1、2、13 的数据的 SQL 语句。修改例 9-12 中的代码，调整后的代码如下所示。

```
1.  <select id="selectDogForEach" resultType="dog">
2.      select * from dog where id in
3.      <foreach collection="idList" open="(" close=")" separator="," item="id">
4.         #{id}
5.      </foreach>
6.  </select>
```

<foreach>元素中的 collection 属性负责引入需要进行循环操作的集合，open 属性负责设置拼接 SQL 语句的前缀，close 属性负责设置拼接 SQL 语句的后缀，separator 属性负责设置每次循环中元素的分隔

符，item 属性代表循环中的每一个元素。除此之外，还有 index 属性，它代表此次循环的角标。

修改 Mapper 接口的代码，调整后的代码如下所示。

```
/*查询*/
List<Dog> selectDogForEach(@Param("idList") ArrayList idList);
```

在此使用@Param 注解标注值的名字，在映射文件中可直接以名字取出相应的值，@Param 注解的内容将在 9.4 节详解。

修改测试类，调整后的代码如下所示。

```
1.  /*通过 SqlSessionFactory 类创建 SqlSession 对象*/
2.  SqlSession sqlSession = build.openSession();
3.  /*获取 MyBatis 提供的实现类*/
4.  DogMapper mapper = sqlSession.getMapper(DogMapper.class);
5.
6.  /*调用查询方法*/
7.  ArrayList<Integer> integer = new ArrayList<>();
8.  integer.add(1);
9.  integer.add(2);
10. integer.add(13);
11.
12. List<Dog> dogs = mapper.selectDogForEach(integer);
13.
14. for (Dog dog : dogs) {
15.    System.out.println(dog);
16. }
17.
18. /*关闭事务*/
19. sqlSession.close();
```

在以上代码中，调用 selectDogForEach()方法，传入 id 集合。执行测试类，结果如图 9.17 所示。

```
Checking to see if class com.mapper.DogMapper matches criteria [is assignable to Object]
Opening JDBC Connection
Created connection 331418503.
Setting autocommit to false on JDBC Connection [com.mysql.cj.jdbc.ConnectionImpl@13c10b87]
==>  Preparing: select * from dog where id in ( ? , ? , ? )
==> Parameters: 1(Integer), 2(Integer), 13(Integer)
<==    Columns: id, name, age
<==        Row: 1, 小红, 40
<==        Row: 2, 旺财, 2
<==        Row: 13, 旺财二号, 3
<==      Total: 3
Dog{id=1, name='小红', age=40}
Dog{id=2, name='旺财', age=2}
Dog{id=13, name='旺财二号', age=3}
Resetting autocommit to true on JDBC Connection [com.mysql.cj.jdbc.ConnectionImpl@13c10b87]
Closing JDBC Connection [com.mysql.cj.jdbc.ConnectionImpl@13c10b87]
Returned connection 331418503 to pool.

Process finished with exit code 0
```

图 9.17　<foreach>元素测试结果

从图 9.17 中可以看出，SQL 语句拼接正常。

9.2.6 <bind>元素

<bind>元素为命名元素，它可以创建一个变量并将其绑定到上下文中，方便后续重复使用，示例

代码如下所示。

```
1.  <select id="selectDogBind" resultType="dog">
2.      <bind name="bindName" value="'%'+name+'%'"/>
3.      SELECT * FROM dog
4.      <if test="age!=null">
5.          /*相当于where name like '%'+#{name}+'%'*/
6.          where name like #{bindName}
7.      </if>
8.  </select>
```

在以上代码中，使用<bind>元素创建了一个变量，其值为'%'+name+'%'。注意，此处的 name 是从 dog 类中获取的参数，并非 name 字符串。设置<bind>元素后，可以在上下文中使用“#{}”引用创建的变量。

9.3 MyBatis 的关联映射

在实际开发中，由于业务需要，数据表之间往往会存在某种关联关系，如一对一、一对多等。当程序操作数据库时，如果被操作的表与其他表相关联，那么处理这些表中的数据时必须要考虑它们之间的关联关系。MyBatis 提供了简洁的映射方法来处理关联关系。本节将详细介绍 MyBatis 关联关系的处理方法，读者需要熟练掌握本节所有内容。

9.3.1 关联关系概述

表与表之间的关联关系可以分为 3 种，分别为一对一、一对多、多对多。

在一对一的关系中，A 表中的一条数据只能与 B 表中的一条数据关联，反过来同样成立，如学生卡与学生之间的关系。

在一对多的关系中，A 表中的一条数据可以与 B 表中的多条数据关联，但是 B 表中的每条数据都只能与 A 表中的一条数据关联。例如班级和学生之间的关系，每个班级关联多个学生，但是每个学生只能关联一个班级。

在多对多的关系中，A 表中的一条数据可以与 B 表中的多条数据关联，B 表中的每条数据可以与 A 表中的多条数据关联。例如老师和学生之间的关系，一个学生可以关联多个老师，一个老师也可以关联多个学生。

后续的小节将使用 3 个独立的示例分别讲解一对一、一对多、多对多关系的处理方式。

9.3.2 一对一级联查询

将一对一的思想应用到程序中，编写一对一示例代码。

1. 创建实体类与数据表

在 src 文件夹下创建 com 文件夹，在 com 文件夹下创建 pojo 文件夹，在 pojo 文件夹下创建两个实体类，分别为学生卡类与学生类。学生实体类中具有学生卡对象，根据实际情况，学生卡属于学生，所以在学生卡实体类中，不添加学生对象。学生卡类与学生类分别如例 9-15、例 9-16 所示（此处省略 toString()、get()、set()和构造方法）。

【例 9-15】 Card.java

```
1.  public class Card {
2.      Integer id;
3.      String cardNumber;
4.  }
```

【例 9-16】 Student.java

```
1.  public class Student {
2.      Integer id;
3.      String name;
4.      Card card;
5.  }
```

创建对应的数据表。学生卡表 card 的创建语句如下所示。

```
1.  CREATE TABLE 'card'  (
2.    'id' int(0) NOT NULL AUTO_INCREMENT,
3.    'card_number' varchar(255) DEFAULT NULL,
4.    'sid' int(0) NULL DEFAULT NULL,
5.    PRIMARY KEY ('id') USING BTREE
6.  )
```

向 card 表中插入数据，代码如下所示。

```
1.  INSERT INTO 'card' VALUES (1, '178283992', 3);
2.  INSERT INTO 'card' VALUES (2, '123624573', 2);
3.  INSERT INTO 'card' VALUES (3, '234122212', 1);
```

学生卡表 card 如图 9.18 所示。

创建学生表，学生表 student 的创建语句如下所示。

```
1.  CREATE TABLE 'student'  (
2.    'id' int(0) NOT NULL AUTO_INCREMENT,
3.    'name' varchar(255) NULL DEFAULT NULL,
4.    'cid' int(0) NULL DEFAULT NULL,
5.    PRIMARY KEY ('id')
6.  )
```

向 student 表中插入数据，代码如下所示。

```
1.  INSERT INTO 'student' VALUES (1, '张三', 3);
2.  INSERT INTO 'student' VALUES (2, '李四', 2);
3.  INSERT INTO 'student' VALUES (3, '王五', 1);
```

学生表 student 如图 9.19 所示。

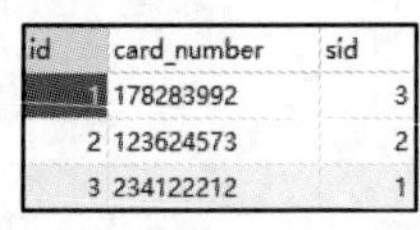

id	card_number	sid
1	178283992	3
2	123624573	2
3	234122212	1

图 9.18　学生卡表 card

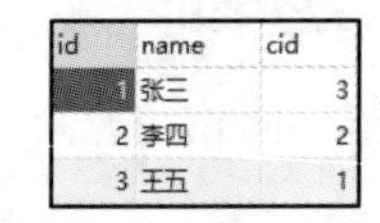

id	name	cid
1	张三	3
2	李四	2
3	王五	1

图 9.19　学生表 student

card 表中拥有学生的 id，根据此 id 即可查询该学生卡的持有人，student 表中每个学生有对应的学生卡的 id，根据此 id 即可查询该学生的学生卡。

2．编写 MyBatis 配置文件与数据库配置文件

在工程文件夹下新建 resource 文件夹，将其标注为资源文件夹。在 resource 文件夹下新建 MyBatis

配置文件，如例 9-17 所示。

【例 9-17】 mybatis-config.xml

```
1.  <?xml version="1.0" encoding="UTF-8" ?>
2.  <!DOCTYPE configuration
3.          PUBLIC "-//mybatis.org//DTD Config 3.0//EN"
4.          "http://mybatis.org/dtd/mybatis-3-config.dtd">
5.  <configuration>
6.      <!--引入配置文件-->
7.      <properties resource="db.properties"/>
8.  
9.      <!--设置 -->
10.     <settings>
11.         <!--开启数据库日志检测-->
12.         <setting name="logImpl" value="STDOUT_LOGGiNG"/>
13.         <!--开启驼峰命名法-->
14.         <setting name="mapUnderscoreToCamelCase" value="true"/>
15.     </settings>
16.     <!--包名简化缩写-->
17.     <typeAliases>
18.         <!--package 方式-->
19.         <package name="com.pojo"/>
20.     </typeAliases>
21.     <!--配置环境-->
22.     <environments default="dev">
23.         <!--配置 MySQL 环境-->
24.         <environment id="dev">
25.             <!--配置事务管理器-->
26.             <transactionManager type="JDBC"/>
27.             <!--配置数据库连接-->
28.             <dataSource type="POOLED">
29.                 <!--配置数据库连接驱动-->
30.                 <property name="driver" value="${jdbc.myDriver}"/>
31.                 <!--配置数据库连接地址-->
32.                 <property name="url" value="${jdbc.myUrl}"/>
33.                 <!--配置用户名-->
34.                 <property name="username" value="${jdbc.myUsername}"/>
35.                 <!--配置密码-->
36.                 <property name="password" value="${jdbc.myPassword}"/>
37.             </dataSource>
38.         </environment>
39.     </environments>
40.     <!--配置 Mapper 映射文件-->
41.     <mappers>
42.         <!--将 com.mapper 包下的所有 Mapper 接口引入-->
43.         <package name="com.mapper"/>
44.     </mappers>
45. </configuration>
```

card 表的 card_number 字段，对应实体类的 cardNumber 属性，这两者之间不能直接进行映射，因此在例 9-17 中的第 14 行代码开启 MyBatis 的驼峰命名法，使其正常映射。此外，pojo 文件夹中存在多个实体类，因此在例 9-17 中的第 19 行代码配置 package 包名缩写，为 com.pojo 文件夹下的所有实体类创建包名缩写。

在 resource 文件夹下创建 db.properties 文件，代码如例 9-18 所示。

【例 9-18】 db.properties

```
1.  jdbc.myDriver = com.mysql.cj.jdbc.Driver
2.  jdbc.myUrl = jdbc:mysql://127.0.0.1/test?\
3.    characterEncoding=utf8&\
4.    useSSL=false&\
5.    serverTimezone=UTC&\
6.    allowPublicKeyRetrieval=true
7.  jdbc.myUsername = root
8.  jdbc.myPassword = root
```

3. 创建映射文件与 Mapper 接口文件

在 com 文件夹下创建 mapper 文件夹，在 mapper 文件夹下新建映射文件与 Mapper 接口文件。编写学生类的映射文件，代码如例 9-19 所示。

【例 9-19】 StudentMapper.xml

```
1.  <?xml version="1.0" encoding="UTF-8"?>
2.  <!DOCTYPE mapper PUBLIC "-//mybatis.org//DTD Mapper 3.0//EN"
3.          "http://mybatis.org/dtd/mybatis-3-mapper.dtd">
4.  <mapper namespace="com.mapper.StudentMapper">
5.      <!--映射结果集-->
6.      <resultMap id="studentMap" type="student">
7.          <id column="id" property="id"/>
8.          <result column="name" property="name"/>
9.          <association property="card" javaType="card">
10.             <id column="cid" property="id"/>
11.             <result column="card_Number" property="cardNumber"/>
12.         </association>
13.     </resultMap>
14.     <!--查询，查询结果使用结果集映射-->
15.     <select id="selectStudent" resultMap="studentMap">
16.         SELECT
17.             s.id,s.name,c.id as cid,c.card_number
18.         FROM
19.             student s
20.         left join
21.             card c
22.         on
23.             s.cid = c.id
24.     </select>
25. </mapper>
```

在例 9-19 中使用连接查询，查询与学生对应的学生卡的信息，并使用结果集来映射，SQL 语句的查询结果如图 9.20 所示。

id	name	cid	card_number
1	张三	3	234122212
2	李四	2	123624573
3	王五	1	178283992

图 9.20 一对一查询结果

在图 9.20 中，查询的结果需要映射到实体类 Student 中。其中，字段 id 和 name 需要映射到 Student 类的 id 和 name 属性中，字段 cid 和 card_number 需要映射到 Student 类中 card 属性对应的 Card 类中。

完成复杂的映射需要使用<resultMap>元素，例 9-19 中的第 6～13 行代码使用了<resultMap>元素。

其中，type 属性是结果集需要映射的对象，id 属性是<resultMap>元素的标识。id 属性在<select>元素的 resultMap 属性中引用。

在<resultMap>元素的子元素中，<id>元素负责映射主键，<result>元素负责映射属性，<association>元素负责映射对象。在这 3 个元素中，column 属性均表示数据库中的字段名，property 属性均表示类中的属性名，在编写代码时，根据映射关系将其完善即可。

在 mapper 文件夹下创建 Mapper 接口文件，代码如例 9-20 所示。

【例 9-20】 StudentMapper.java

```
1.  public interface StudentMapper {
2.      /*查询*/
3.      List<Student> selectStudent();
4.  }
```

4．创建测试类

在 com 文件夹下创建测试类，代码如例 9-21 所示。

【例 9-21】 Test.java

```
1.  public class Test {
2.      public static void main(String[] args) {
3.          /*创建输入流*/
4.          InputStream inputStream = null;
5.
6.          /*将 MyBatis 配置文件转化为输入流*/
7.          try {
8.              inputStream =
9.                      Resources.getResourceAsStream("mybatis-config.xml");
10.         } catch (IOException e) {
11.             e.printStackTrace();
12.         }
13.
14.         /*通过 SqlSessionFactoryBuilder 类创建 SqlSessionFactory 对象*/
15.         SqlSessionFactory build =
16.                 new SqlSessionFactoryBuilder().build(inputStream);
17.
18.         /*通过 SqlSessionFactory 类创建 SqlSession 对象*/
19.         SqlSession sqlSession = build.openSession();
20.         /*获取 MyBatis 提供的实现类*/
21.         StudentMapper mapper = sqlSession.getMapper(StudentMapper.class);
22.
23.         List<Student> students = mapper.selectStudent();
24.
25.         for (Student student : students) {
26.             System.out.println(student);
27.         }
28.
29.         /*关闭事务*/
30.         sqlSession.close();
31.     }
32. }
```

执行例 9-21 中的代码，结果如图 9.21 所示。

```
Checking to see if class com.mapper.StudentMapper matches criteria [is assignable to Object]
Opening JDBC Connection
Created connection 898406901.
Setting autocommit to false on JDBC Connection [com.mysql.cj.jdbc.ConnectionImpl@358c99f5]
==>  Preparing: SELECT s.id,s.name,c.id as cid,c.card_number FROM student s left join card c on s.cid = c.id
==> Parameters:
<==    Columns: id, name, cid, card_number
<==        Row: 1, 张三, 3, 234122212
<==        Row: 2, 李四, 2, 123624573
<==        Row: 3, 王五, 1, 178283992
<==      Total: 3
Student{id=1, name='张三', card=Card{id=3, cardNumber='234122212'}}
Student{id=2, name='李四', card=Card{id=2, cardNumber='123624573'}}
Student{id=3, name='王五', card=Card{id=1, cardNumber='178283992'}}
Resetting autocommit to true on JDBC Connection [com.mysql.cj.jdbc.ConnectionImpl@358c99f5]
Closing JDBC Connection [com.mysql.cj.jdbc.ConnectionImpl@358c99f5]
Returned connection 898406901 to pool.

Process finished with exit code 0
```

图 9.21　一对一测试结果

从图 9.21 中可以看出，card 对象中的属性映射成功。

9.3.3 一对多级联查询

将一对多的思想应用到程序中，编写一对多示例代码。

1. 创建实体类与数据表

参考 9.3.2 小节的内容创建目录结构，在 pojo 文件夹内创建两个实体类，分别为班级类与学生类。班级实体类中存在学生对象，因为班级与学生之间为从属关系，所以学生实体类中不添加班级类。班级类与学生类分别如例 9-22、例 9-23 所示（此处省略 toString()、get()、set()和构造方法）。

【例 9-22】 MyClass.java

```
1.  public class MyClass {
2.      Integer id;
3.      String className;
4.      List<Student> student;
5.  }
```

【例 9-23】 Student.java

```
1.  public class Student {
2.      Integer id;
3.      String name;
4.  }
```

创建对应的数据表。班级表 myclass 的创建语句如下所示。

```
1.  CREATE TABLE 'myclass'  (
2.    'id' int(0) NOT NULL AUTO_INCREMENT,
3.    'class_name' varchar(255) NULL DEFAULT NULL,
4.    PRIMARY KEY ('id')
5.  )
```

向 myclass 表中插入数据，SQL 代码如下所示。

```
1.  INSERT INTO 'myclass' VALUES (1, '一班');
2.  INSERT INTO 'myclass' VALUES (2, '二班');
3.  INSERT INTO 'myclass' VALUES (3, '三班');
```

班级表 myclass 如图 9.22 所示。

创建学生表，学生表 student 的创建语句如下所示。

```
1.  CREATE TABLE 'student'  (
2.    'id' int(0) NOT NULL AUTO_INCREMENT,
3.    'name' varchar(255) NULL DEFAULT NULL,
4.    'cid' int(0),
5.    PRIMARY KEY ('id')
6.  )
```

向 student 表中插入数据，SQL 代码如下所示。

```
1.  INSERT INTO 'student' VALUES (1, '张三', 1);
2.  INSERT INTO 'student' VALUES (2, '李四', 2);
3.  INSERT INTO 'student' VALUES (3, '王五', 1);
4.  INSERT INTO 'student' VALUES (4, '赵六', 3);
5.  INSERT INTO 'student' VALUES (5, '蜀七', 1);
6.  INSERT INTO 'student' VALUES (6, '魏八', 3);
```

学生表 student 如图 9.23 所示。

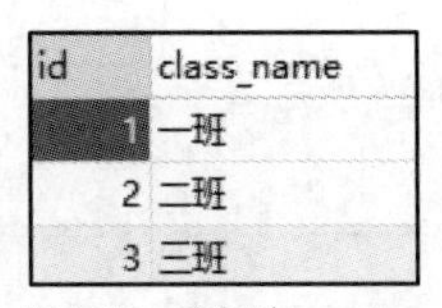

id	class_name
1	一班
2	二班
3	三班

图 9.22　班级表 myclass

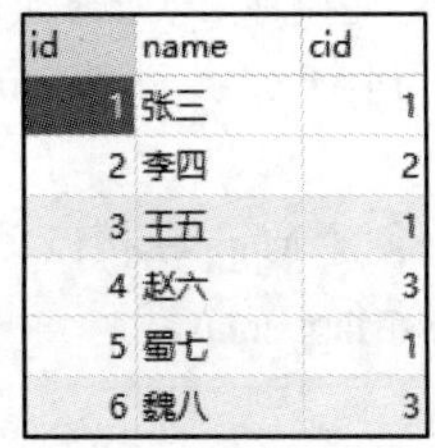

id	name	cid
1	张三	1
2	李四	2
3	王五	1
4	赵六	3
5	蜀七	1
6	魏八	3

图 9.23　学生表 student

在 student 表中每个学生有对应的班级 id，根据此 id 即可查询该学生所在的班级。

2．编写 MyBatis 配置文件与数据库配置文件

MyBatis 配置文件如例 9-17 所示，数据库配置文件如例 9-18 所示。

3．创建映射文件与 Mapper 接口文件

在 mapper 文件夹中新建映射文件与 Mapper 接口文件。编写班级类的映射文件，代码如例 9-24 所示。

【例 9-24】 MyClassMapper.xml

```
1.  <?xml version="1.0" encoding="UTF-8"?>
2.  <!DOCTYPE mapper PUBLIC "-//mybatis.org//DTD Mapper 3.0//EN"
3.          "http://mybatis.org/dtd/mybatis-3-mapper.dtd">
4.  <mapper namespace="com.mapper.MyClassMapper">
5.      <!--映射结果集-->
6.      <resultMap id="myClassMap" type="myClass">
7.          <id column="cid" property="id"/>
8.          <result column="class_name" property="className"/>
9.          <collection property="student" ofType="student">
10.             <id column="id" property="sid"/>
11.             <result column="name" property="name"/>
12.         </collection>
13.     </resultMap>
14.
```

```
15.     <!--查询，查询结果使用结果集映射-->
16.     <select id="selectMyClass" resultMap="myClassMap">
17.         SELECT
18.             c.id cid,class_name,s.id sid,s.name
19.         FROM
20.             myclass c
21.         left join
22.              student s
23.         on
24.             c.id = s.cid
25.     </select>
26. </mapper>
```

在例 9-24 中使用连接查询来查询班级的信息，并使用结果集映射，利用<collection>元素来填充 student 集合的属性。在<collection>元素中，ofType 属性表示集合中的对象类型。例 9-24 中 SQL 语句的查询结果如图 9.24 所示。

cid	class_name	sid	name
1	一班	5	蜀七
1	一班	3	王五
1	一班	1	张三
2	二班	2	李四
3	三班	6	魏八
3	三班	4	赵六

图 9.24　一对多查询结果

在图 9.24 中，查询的结果需要映射到实体类 MyClass 中。其中，字段 cid 和 class_name 需要映射到 MyClass 类的 id 和 className 属性中；字段 sid 和 name 需要映射到 MyClass 类中 student 属性对应的 Student 集合中，分别对应 Student 对象的 id 属性和 name 属性。

在 mapper 文件夹下创建 Mapper 接口文件，代码如例 9-25 所示。

【例 9-25】 MyClassMapper.java

```
1.  public interface MyClassMapper {
2.      /*查询*/
3.      List<MyClass> selectMyClass();
4.  }
```

4．创建测试类

在 com 文件夹下创建测试类，代码如例 9-26 所示。

【例 9-26】 Test.java

```
1.  public class Test {
2.      public static void main(String[] args) {
3.          /*创建输入流*/
4.          InputStream inputStream = null;
5.
6.          /*将 MyBatis 配置文件转化为输入流*/
7.          try {
8.              inputStream =
9.                      Resources.getResourceAsStream("mybatis-config.xml");
10.         } catch (IOException e) {
11.             e.printStackTrace();
12.         }
13.
14.         /*通过 SqlSessionFactoryBuilder 类创建 SqlSessionFactory 对象*/
15.         SqlSessionFactory build =
16.                 new SqlSessionFactoryBuilder().build(inputStream);
17.
18.         /*通过 SqlSessionFactory 类创建 SqlSession 对象*/
```

```
19.         SqlSession sqlSession = build.openSession();
20.         /*获取 MyBatis 提供的实现类*/
21.         MyClassMapper mapper = sqlSession.getMapper(MyClassMapper.class);
22.
23.         List<MyClass> myClasses = mapper.selectMyClass();
24.
25.         for (MyClass myClass : myClasses) {
26.             System.out.println(myClass);
27.         }
28.
29.         /*关闭事务*/
30.         sqlSession.close();
31.     }
32. }
```

执行例 9-26 中的代码，结果如图 9.25 所示。

```
Checking to see if class com.mapper.MyClassMapper matches criteria [is assignable to Object]
Opening JDBC Connection
Created connection 898406901.
Setting autocommit to false on JDBC Connection [com.mysql.cj.jdbc.ConnectionImpl@358c99f5]
==>  Preparing: SELECT c.id cid,class_name,s.id sid,s.name FROM myclass c left join student s on c.id = s.cid
==> Parameters: 
<==    Columns: cid, class_name, sid, name
<==        Row: 1, 一班, 5, 蜀七
<==        Row: 1, 一班, 3, 王五
<==        Row: 1, 一班, 1, 张三
<==        Row: 2, 二班, 2, 李四
<==        Row: 3, 三班, 6, 魏八
<==        Row: 3, 三班, 4, 赵六
<==      Total: 6
MyClass{id=1, className='一班', student=[Student{id=null, name='蜀七'}, Student{id=null, name='王五'}, Student{id=null, name='张三'}]}
MyClass{id=2, className='二班', student=[Student{id=null, name='李四'}]}
MyClass{id=3, className='三班', student=[Student{id=null, name='魏八'}, Student{id=null, name='赵六'}]}
Resetting autocommit to true on JDBC Connection [com.mysql.cj.jdbc.ConnectionImpl@358c99f5]
Closing JDBC Connection [com.mysql.cj.jdbc.ConnectionImpl@358c99f5]
Returned connection 898406901 to pool.

Process finished with exit code 0
```

图 9.25 一对多查询测试结果

从图 9.25 中可以看出，Student 集合对象封装完成。

9.3.4 多对多级联查询

将多对多的思想应用到程序中，编写多对多示例代码。

1. 创建实体类与数据表

在 pojo 文件夹下创建两个实体类，分别为学生类与教师类，学生实体类中具有教师集合，教师实体类中也有学生集合。学生类与教师类分别如例 9-27、例 9-28 所示（此处省略 toString()、get()、set() 和构造方法）。

【例 9-27】 Student.java

```
1.  public class Student {
2.      Integer id;
3.      String name;
4.      List<Teacher> teacher;
5.  }
```

【例 9-28】 Teacher.java

```
1.  public class Teacher {
2.      Integer id;
```

```
3.      String courseName;
4.      String name;
5.      List<Student> student;
6.  }
```

创建对应的数据表。因为两个表之间的关系为多对多，所以需要将多对多拆分成多个一对多，通过中间表进行连接，利用连接查询可解决级联查询问题。

学生表 student 的创建语句如下所示。

```
1.  CREATE TABLE 'student'  (
2.    'id' int(0) NOT NULL AUTO_INCREMENT,
3.    'name' varchar(255) NULL DEFAULT NULL,
4.    PRIMARY KEY ('id')
5.  )
```

向 student 表中插入数据，代码如下所示。

```
1.  INSERT INTO 'student' VALUES (1, '张三');
2.  INSERT INTO 'student' VALUES (2, '李四');
3.  INSERT INTO 'student' VALUES (3, '王五');
```

学生表 student 如图 9.26 所示。

创建教师表，教师表 teacher 的创建语句如下所示。

```
1.  CREATE TABLE 'teacher'  (
2.    'id' int(0) NOT NULL,
3.    'course_name' varchar(255)  NULL DEFAULT NULL,
4.    'name' varchar(255)  NULL DEFAULT NULL,
5.    PRIMARY KEY ('id')
6.  )
```

向 teacher 表中插入数据，代码如下所示。

```
1.  INSERT INTO 'teacher' VALUES (1, '语文', '刘伟');
2.  INSERT INTO 'teacher' VALUES (2, '语文', '张帅');
3.  INSERT INTO 'teacher' VALUES (3, '数学', '赵凯');
4.  INSERT INTO 'teacher' VALUES (4, '英语', '刘波');
5.  INSERT INTO 'teacher' VALUES (5, '英语', '英丽');
```

教师表 teacher 如图 9.27 所示。

id	name
1	张三
2	李四
3	王五

图 9.26　学生表 student

id	course_name	name
1	语文	刘伟
2	语文	张帅
3	数学	赵凯
4	英语	刘波
5	英语	英丽

图 9.27　教师表 teacher

创建中间表，中间表 student_teacher 的创建语句如下所示。

```
1.  CREATE TABLE 'student_teacher'  (
2.    'id' int(0) NOT NULL AUTO_INCREMENT,
3.    't_id' int(0) NULL DEFAULT NULL,
4.    's_id' int(0) NULL DEFAULT NULL,
5.    PRIMARY KEY ('id')
6.  )
```

向 student_teacher 表中插入数据，代码如下所示。

```
1.  INSERT INTO 'student_teacher' VALUES (1, 1, 1);
2.  INSERT INTO 'student_teacher' VALUES (2, 3, 1);
3.  INSERT INTO 'student_teacher' VALUES (3, 5, 1);
4.  INSERT INTO 'student_teacher' VALUES (4, 2, 2);
5.  INSERT INTO 'student_teacher' VALUES (5, 3, 2);
6.  INSERT INTO 'student_teacher' VALUES (6, 4, 2);
7.  INSERT INTO 'student_teacher' VALUES (7, 1, 3);
8.  INSERT INTO 'student_teacher' VALUES (8, 3, 3);
9.  INSERT INTO 'student_teacher' VALUES (9, 4, 3);
```

中间表 student_teacher 表如图 9.28 所示。

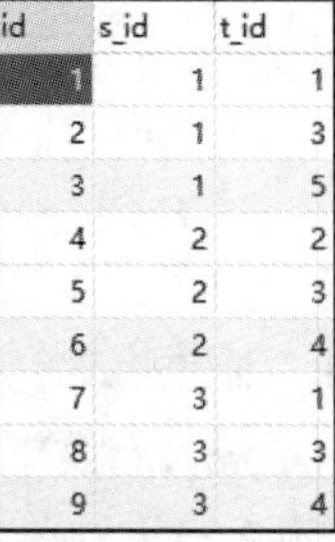

id	s_id	t_id
1	1	1
2	1	3
3	1	5
4	2	2
5	2	3
6	2	4
7	3	1
8	3	3
9	3	4

图 9.28 中间表 student_teacher

2．编写 MyBatis 配置文件与数据库配置文件

MyBatis 配置文件如例 9-17 所示，数据库配置文件如例 9-18 所示。

3．创建映射文件与 Mapper 接口文件

在 mapper 文件夹中新建映射文件与 Mapper 接口文件。编写学生类的映射文件，代码如例 9-29 所示。

【例 9-29】 StudentMapper.xml

```
1.  <?xml version="1.0" encoding="UTF-8"?>
2.  <!DOCTYPE mapper PUBLIC "-//mybatis.org//DTD Mapper 3.0//EN"
3.          "http://mybatis.org/dtd/mybatis-3-mapper.dtd">
4.  <mapper namespace="com.mapper.StudentMapper">
5.      <!--映射结果集-->
6.      <resultMap id="studentMap" type="student">
7.          <id column="sid" property="id"/>
8.          <result column="sname" property="name"/>
9.          <collection property="teacher" ofType="teacher">
10.             <id column="id" property="tid"/>
11.             <result column="course_name" property="courseName"/>
12.             <result column="tname" property="name"/>
13.         </collection>
14.     </resultMap>
15.
16.     <!--查询，查询结果使用结果集映射-->
17.     <select id="selectStudent" resultMap="studentMap">
18.         SELECT
19.             s.id sid,s.name sname,t.id tid,t.course_name,t.name tname
20.         FROM
21.             student s
22.         join
23.             student_teacher st
24.         on
25.             s.id=st.s_id
26.         join
27.             teacher t
28.         on
29.             st.t_id = t.id
30.     </select>
31. </mapper>
```

在例 9-29 中使用连接查询来查询学生的信息，并使用结果集映射，SQL 语句的查询结果如图 9.29 所示。

在图 9.29 中，查询的结果需要映射到实体类 Student 中。其中，字段 sid 和 sname 需要映射到 Student 类的 id 和 name 属性中；字段 tid、course_name 和 tname 需要映射到 Student 类中的 Teacher 集合中，分别对应 id、courseName 和 name。

在 mapper 文件夹下创建 Mapper 接口文件，代码如例 9-30 所示。

sid	sname	tid	course_name	tname
1	张三	1	语文	刘伟
1	张三	3	数学	赵凯
1	张三	5	英语	英丽
2	李四	2	语文	张帅
2	李四	3	数学	赵凯
2	李四	4	英语	刘波
3	王五	1	语文	刘伟
3	王五	3	数学	赵凯
3	王五	4	英语	刘波

图 9.29　多对多查询结果

【例 9-30】 StudentMapper.java

```
1.  public interface StudentMapper {
2.      /*查询*/
3.      List<Student> selectStudent();
4.  }
```

4．创建测试类

在 com 文件夹下创建测试类，代码如例 9-31 所示。

【例 9-31】 Test.java

```
1.  public class Test {
2.      public static void main(String[] args) {
3.          /*创建输入流*/
4.          InputStream inputStream = null;
5.
6.          /*将 MyBatis 配置文件转化为输入流*/
7.          try {
8.              inputStream =
9.                      Resources.getResourceAsStream("mybatis-config.xml");
10.         } catch (IOException e) {
11.             e.printStackTrace();
12.         }
13.
14.         /*通过 SqlSessionFactoryBuilder 类创建 SqlSessionFactory 对象*/
15.         SqlSessionFactory build =
16.                 new SqlSessionFactoryBuilder().build(inputStream);
17.
18.         /*通过 SqlSessionFactory 类创建 SqlSession 对象*/
19.         SqlSession sqlSession = build.openSession();
20.         /*获取 MyBatis 提供的实现类*/
21.         StudentMapper mapper = sqlSession.getMapper(StudentMapper.class);
22.
23.         List<Student> students = mapper.selectStudent();
24.
25.         for (Student student : students) {
26.             System.out.println(student);
27.         }
28.
29.         /*关闭事务*/
30.         sqlSession.close();
31.     }
32. }
```

执行例 9-31 中的代码，结果如图 9.30 所示。

```
Checking to see if class com.mapper.StudentMapper matches criteria [is assignable to Object]
Opening JDBC Connection
Created connection 898406901.
Setting autocommit to false on JDBC Connection [com.mysql.cj.jdbc.ConnectionImpl@358c99f5]
==>  Preparing: SELECT s.id sid,s.name sname,t.id tid,t.course_name,t.name tname FROM student s join student_teacher st on s.id=st.s_id join teacher t on st.t_id = t.id
==> Parameters: 
<==    Columns: sid, sname, tid, course_name, tname
<==        Row: 1, 张三, 1, 语文, 刘伟
<==        Row: 1, 张三, 3, 数学, 赵凯
<==        Row: 1, 张三, 5, 英语, 英丽
<==        Row: 2, 李四, 2, 语文, 张帅
<==        Row: 2, 李四, 3, 数学, 赵凯
<==        Row: 2, 李四, 4, 英语, 刘波
<==        Row: 3, 王五, 1, 语文, 刘伟
<==        Row: 3, 王五, 3, 数学, 赵凯
<==        Row: 3, 王五, 4, 英语, 刘波
<==      Total: 9
Student{id=1, name='张三', teacher=[Teacher{id=null, courseName='语文', name='刘伟', student=null}, Teacher{id=null, courseName='数学', name='赵凯', student=null}, Teacher{id=null, courseName='英语', name='英丽', student=null}]}
Student{id=2, name='李四', teacher=[Teacher{id=null, courseName='语文', name='张帅', student=null}, Teacher{id=null, courseName='数学', name='赵凯', student=null}, Teacher{id=null, courseName='英语', name='刘波', student=null}]}
Student{id=3, name='王五', teacher=[Teacher{id=null, courseName='语文', name='刘伟', student=null}, Teacher{id=null, courseName='数学', name='赵凯', student=null}, Teacher{id=null, courseName='英语', name='刘波', student=null}]}
Resetting autocommit to true on JDBC Connection [com.mysql.cj.jdbc.ConnectionImpl@358c99f5]
Closing JDBC Connection [com.mysql.cj.jdbc.ConnectionImpl@358c99f5]
Returned connection 898406901 to pool.

Process finished with exit code 0
```

图 9.30　多对多查询测试结果

多对多查询与一对多查询的区别不大。相对于一对多查询，多对多查询需要用到较多的连接，在实际开发时需要着重注意 SQL 语句的编写。

9.4 MyBatis 的注解开发

在 MyBatis 中，除了通过 XML 映射文件创建 SQL 语句，还可以通过注解直接编写 SQL 语句，此方式在开发中经常使用，读者需要熟练掌握。

9.4.1 注解开发简介

在使用注解开发时，无须创建映射文件，直接在 Mapper 接口文件中通过注解编写 SQL 语句即可。MyBatis 提供了若干个注解来支持注解开发，常用的注解如表 9.3 所示。

表 9.3　MyBatis 注解开发中的常用注解

注解	说明
@Select	查询操作注解
@Insert	插入操作注解
@Update	更新操作注解
@Delete	删除操作注解
@Param	用于标注传入参数的名称

9.4.2 注解开发的简单应用

本示例中的实体类采用例 9-23 的 Student 类。配置文件与数据库连接文件分别采用例 9-17、例 9-18 的文件。在 mapper 文件夹下创建 StudentMapper 接口文件，如例 9-32 所示。

【例 9-32】 StudentMapper.java

```
1.  public interface StudentMapper {
2.
3.      /*查询*/
4.      @Select("select * from student")
5.      List<Student> getAllStudent();
```

```
6.
7.      /*通过指定参数查询学生信息，属性 testId 可以任意命名*/
8.      @Select("select * from student where id=#{id}")
9.      Student getStudentById(@Param("id") int testId);
10.
11.     /*插入学生信息*/
12.     @Insert("insert into student(name) values (#{name})")
13.     int addStudent(Student student);
14.
15.     /*修改学生信息*/
16.     @Update("update student set name=#{name} where id=#{id}")
17.     int updateStudent(@Param("id") int id,@Param("name") String name);
18.
19.     /*删除学生信息*/
20.     @Delete("delete from student where id=#{id}")
21.     int deleteStudent(@Param("id") int id);
22.
23. }
```

下面对例 9-32 进行详细讲解。

在注解开发中，将注解的语句直接放在 Mapper 接口文件的方法上方，代表此方法对应的操作。在第 4 行代码中，使用@Select 注解完成具有查询功能的编写，查询的 SQL 语句直接写在注解后的括号中，其他注解（如@Update、@Insert 和@Delete 注解）的使用方法与@Select 注解相同。

在编写注解中的 SQL 语句时，同样可以使用“#{}”来取值，如第 8 行代码。

向 SQL 语句中传参的方式有三种：第一种方式为传入对象，SQL 语句通过对象中的属性名取值；第二种方式为传入 Map 集合，SQL 语句通过 Map 集合中的键来取值；第三种方式为在 Mapper 接口的方法中使用@Param 注解标注一个参数，如第 9 行代码，此@Param 注解代表此参数的名字，SQL 语句可以通过@Param 注解中的参数名取出对应的值。

编写测试类，代码如例 9-33 所示。

【例 9-33】 Test.java

```
1.  public class Test {
2.      public static void main(String[] args) {
3.          /*创建输入流*/
4.          InputStream inputStream = null;
5.
6.          /*将 MyBatis 配置文件转化为输入流*/
7.          try {
8.              inputStream =
9.                      Resources.getResourceAsStream("mybatis-config.xml");
10.         } catch (IOException e) {
11.             e.printStackTrace();
12.         }
13.
14.         /*通过 SqlSessionFactoryBuilder 类创建 SqlSessionFactory 对象*/
15.         SqlSessionFactory build =
16.                 new SqlSessionFactoryBuilder().build(inputStream);
17.
18.         /*通过 SqlSessionFactory 类创建 SqlSession 对象*/
19.         SqlSession sqlSession = build.openSession();
```

```
20.         /*获取MyBatis 提供的实现类*/
21.         StudentMapper mapper =
22.                 sqlSession.getMapper(StudentMapper.class);
23.
24.         /*测试查询的方法*/
25.         List<Student> students = mapper.getAllStudent();
26.
27.         /*测试根据 id 查询*/
28.         Student studentById = mapper.getStudentById(1);
29.
30.         /*测试添加*/
31.         int insertDex = mapper.addStudent(new Student( "我是新来的"));
32.
33.         /*测试更新*/
34.         int updateDex = mapper.updateStudent(1,"更改过后的名字");
35.
36.         /*测试删除*/
37.         int deleteDex = mapper.deleteStudent(3);
38.
39.         /*测试全部查询*/
40.         for (Student student : students) {
41.             System.out.println("测试全部查询:"+ student);
42.         }
43.         /*测试根据 id 查询*/
44.         System.out.println("测试根据 id 查询:"+studentById);
45.         /*测试添加*/
46.         if (insertDex == 1){
47.             System.out.println("测试插入:插入成功");
48.         }else{
49.             System.out.println("测试插入:插入失败");
50.         }
51.         /*测试更新*/
52.         if (updateDex == 1){
53.             System.out.println("测试更新:更新成功");
54.         }else{
55.             System.out.println("测试更新:更新失败");
56.         }
57.         /*测试删除*/
58.         if (deleteDex == 1){
59.             System.out.println("测试删除:删除成功");
60.         }else{
61.             System.out.println("测试删除:删除失败");
62.         }
63.         /*提交事务*/
64.         sqlSession.commit();
65.         /*关闭事务*/
66.         sqlSession.close();
67.     }
68. }
```

执行例 9-33 中的测试类，输出结果如图 9.31 所示。

```
Setting autocommit to false on JDBC Connection [com.mysql.cj.jdbc.ConnectionImpl@51931956]
==>  Preparing: select * from student
==> Parameters: 
<==    Columns: id, name
<==        Row: 1, 张三
<==        Row: 2, 李四
<==        Row: 3, 王五
<==      Total: 3
==>  Preparing: select * from student where id=?
==> Parameters: 1(Integer)
<==    Columns: id, name
<==        Row: 1, 张三
<==      Total: 1
==>  Preparing: insert into student(name) values (?)
==> Parameters: 我是新来的(String)
<==    Updates: 1
==>  Preparing: update student set name=? where id=?
==> Parameters: 更改过后的名字(String), 1(Integer)
<==    Updates: 1
==>  Preparing: delete from student where id=?
==> Parameters: 3(Integer)
<==    Updates: 1
测试全部查询:Student{id=1, name='张三'}
测试全部查询:Student{id=2, name='李四'}
测试全部查询:Student{id=3, name='王五'}
测试根据id查询:Student{id=1, name='张三'}
测试插入:插入成功
测试更新:更新成功
测试删除:删除成功
Committing JDBC Connection [com.mysql.cj.jdbc.ConnectionImpl@51931956]
Resetting autocommit to true on JDBC Connection [com.mysql.cj.jdbc.ConnectionImpl@51931956]
Closing JDBC Connection [com.mysql.cj.jdbc.ConnectionImpl@51931956]
Returned connection 1368594774 to pool.

Process finished with exit code 0
```

图 9.31　MyBatis 注解开发的测试结果

从图 9.31 中可以看出，通过注解成功进行了数据库中数据的增删改查操作。

在此需要注意的是，MyBatis 注解开发除了可进行简单的增删改查操作外，还可进行一对多、多对多等级联查询。因为一对多、多对多等级联查询需要复杂的配置，一般使用 XML 映射文件编写，所以在此不再介绍使用注解方式实现级联查询的方法，感兴趣的读者可以自行参考相关书籍。

在实际开发中，当 SQL 语句比较复杂时，一般将注解方式与 XML 方式混合使用，复杂的 SQL 语句使用 XML 映射文件编写，简单的 SQL 语句使用注解编写。

9.5 本章小结

本章首先介绍了 MyBatis 的缓存机制，然后介绍了 MyBatis 动态 SQL 的使用方法，最后讲解了 MyBatis 的级联查询与注解开发，其中，动态 SQL 和 MyBatis 关联映射是本章的重点内容。

在本章内容中，MyBatis 缓存是需要理解的内容，在开发中使用不多，在面试时有可能提到。动态 SQL 和 MyBatis 关联映射是开发中常用的内容，需要重点掌握。MyBatis 注解开发在简单 SQL 语句中较常用，复杂 SQL 语句一般使用 XML 文件编写。

9.6 习题

1．填空题

（1）MyBatis 的一级缓存存放在________执行器中。

（2）MyBatis 在执行到________语句时，会清空当前命名空间中所有的缓存。

（3）在 MyBatis 动态 SQL 中，负责循环拼接 SQL 语句的元素是________。

（4）在 MyBatis 的多对多级联查询中，利用映射结果集元素________进行对象的映射。

2．选择题

（1）关于 MyBatis 的缓存，下列描述错误的是（　　）。

A. 当开启 MyBatis 二级缓存时，需要注意跨命名空间的操作是否会引起脏读问题

B. 查询语句之前进行 DML 操作，则此后查询语句不会取用缓存中的数据

C. 在 MyBatis 二级缓存关闭的状态下，在两次请求中，如果 SQL 查询语句完全相同，则对于第二次的请求 SQL 语句将使用一级缓存中的内容

D. 在执行两次连续的查询 SQL 时，参数不同，但最终 SQL 语句相同的情况下，这两次查询操作仍然不能认定为相同

（2）下列关于动态 SQL 的表述错误的是（　　）。

A. <foreach>元素可以支持数组、List 集合和 Set 集合的遍历

B. <if>元素负责判断某条 SQL 语句的拼接是否符合条件

C. <trim>元素可以模拟<where>和<set>元素的功能，完成 SQL 语句的拼接

D. <bind>元素可以绑定某条 SQL 语句，使其在满足相应条件时生效

（3）下列关于 MyBatis 注解的表述错误的是（　　）。

A. @Select 注解用于映射查询语句，其作用等同于 XML 文件中的<select>元素

B. @Update 注解用于映射插入语句，其作用等同于 XML 文件中的<update>元素

C. @Delete 注解用于映射删除语句，其作用等同于 XML 文件中的<delete>元素

D. @Param 注解用于指定传入参数的名称，通常用于非对象、非 Map 集合的直接属性的传递

3．思考题

（1）简述 MyBatis 的缓存机制。

（2）简述 MyBatis 动态 SQL 的使用方法。

4．编程题

（1）创建 Dog 类，在其中添加 name 和 age 属性。创建对应数据库，根据 name 和 age 属性，使用 MyBatis 查询和更新 Dog 类，要求当 name 和 age 属性的值为空时，去掉相应的筛选条件（使用动态 SQL 实现）。

（2）在第（1）题的基础上，查询 id 为 3、8 和 2 的数据，要求使用动态 SQL 实现。

（3）创建 Company 类和 Employee 类，在 Company 类中添加 Employee 属性和 name 属性，在 Employee 类中添加 name 属性，在数据库创建 Company 表和 Employee 表，通过 MyBatis 查询 id 为 1 的 Company 类的信息（Company 类中需要含有所属的 Employee 类的列表）。

第10章 Spring MVC

本章学习目标

- 理解 MVC 模式。
- 理解 Spring MVC 的核心组件。
- 掌握 Spring MVC 的简单应用。
- 掌握 Spring MVC 常用注解的使用方法。

在 Web 企业级开发中，MVC 是构建 Web 程序的常用架构。在 MVC 模式中，M 表示 Model（模型），V 表示 View（视图），C 表示 Controller（控制器）。Spring MVC 就是以 MVC 模式为基础构建的框架，它负责接收客户端的请求，并将其转发给业务层处理，同时也负责将处理后的结果返回给客户端。本章将讲解 Spring MVC 的基础知识。

10.1 Spring MVC 概述

Spring MVC 是 Web 框架的控制器。本节主要讲解 Spring MVC 的功能与 MVC 模式，读者需要初步了解 Spring MVC 的功能，熟悉 MVC 模式。

10.1.1 Spring MVC 简介

Spring MVC 是 Spring 框架中的一个模块，同时也是一个可用于构建 Web 程序的 MVC 框架。它提供了一个前端控制器来发送请求，同时，它还支持包括 JSP、FreeMarker、Velocity 等在内的多项视图技术。

Spring MVC 灵活、高效，且配置方便。与其他 MVC 框架相比，它可以与 Spring 无缝集成，直接使用 Spring 的功能，具有良好的复用性与扩展性。

10.1.2 MVC 模式

MVC 模式是一种开发模式，它把整个 Web 访问的流程分为 4 个层级，分别为 Controller 层、Service 层、DAO 层（又称 Mapper 层）和 POJO 层。MVC 的结构如图 10.1 所示。

从图 10.1 中可以看出，MVC 可以拆分为 Model、View 和 Controller 这 3 个部分，其中 Model 表示模型，在代码中体现为 DAO 层与 Service 层；View 表示视图，即最终展示给用户的结果；Controller 表示控制器，负责处理和转发用户的请求。下面以 MVC 开发模式来规范日常 Web 开发时的目录结构，如图 10.2 所示。

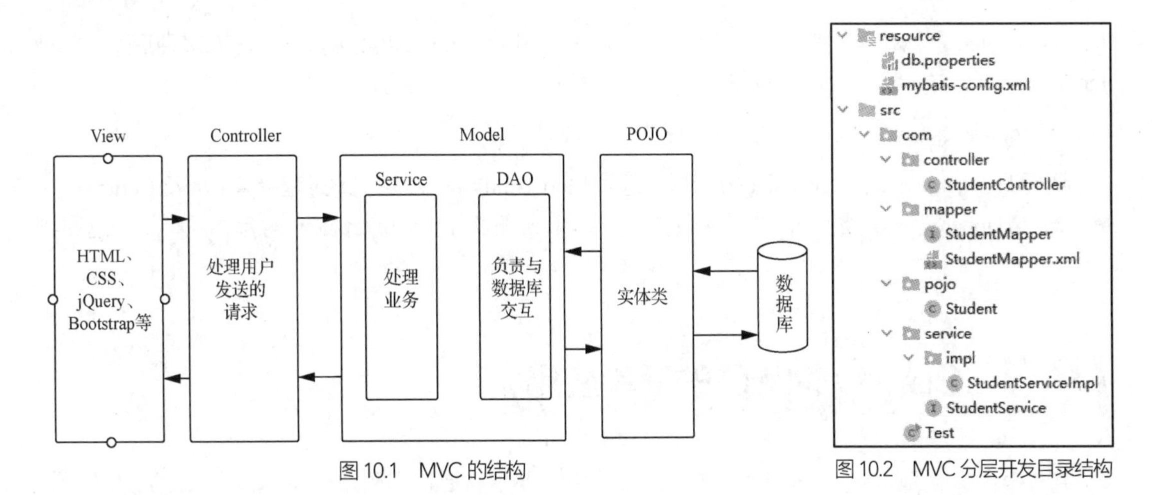

图 10.1 MVC 的结构

图 10.2 MVC 分层开发目录结构

下面详细介绍图 10.2 中的目录结构。

（1）resource 文件夹中存放 MyBatis 配置文件与数据库配置文件。

（2）controller 文件夹中存放控制器层的类，控制器将请求发送给 Service 层进行业务处理。

（3）service 文件夹中存放业务方法，因为业务方法过多时不易于调试与维护，所以此处将具体的实现类放在 impl 文件夹中，将实现类的所有功能抽取到接口文件中，并将该文件存放于 service 文件夹中。需要注意的是，impl 文件夹中的类必须实现 service 文件夹中对应的接口。

（4）mapper 文件夹中存放的是与数据库交互的 Mapper 接口文件与映射文件。

（5）pojo 文件夹中存放的是实体类。

10.2 Spring MVC 的核心组件

Spring 通过一系列组件来实现 Spring MVC 的功能，这些组件包括 DispatcherServlet（前端控制器）、HandlerMapping（处理器映射器）、Handler（处理器）、HandlerAdapter（处理器适配器）和 ViewResolver（视图解析器）等。本节将从这些组件入手，带领读者了解 Spring MVC 的执行流程。

1．DispatcherServlet

DispatcherServlet 负责拦截客户端的请求并将其分发给其他组件。在 Web 开发中，DispatcherServlet 是 Spring MVC 整个执行流程的控制中枢。

2．HandlerMapping

HandlerMapping 是一个 Map 集合，用于处理 URL 和处理类之间的映射，其中键是传入请求的 URL，值是处理类的全限定类名。当请求传入 DispatcherServlet 中时，Spring MVC 使用 HandlerMapping 对 URL 进行映射，根据 URL 找到对应的 Handler。

3．Handler

Handler 是一个负责处理请求的方法，它可以是一个 Controller，也可以是一个 HandlerMethod（其中封装了 Controller 的方法）。

4．HandlerAdapter

使用 HandlerMapping 查找到对应的 Handler 后，根据此 Handler 获取相应的 HandlerAdapter。

HandlerAdapter 是采用适配器设计模式设计的适配类，因为 Spring MVC 的 Handler 有多种形态，每种 Handler 的调用方式不同，所以需要使用适配器适配 Handler 与处理方法。

5．ViewResolver

调用 HandlerAdapter 的 handler()方法后，返回 ModelAndView 对象（此过程中可能涉及 Controller、Service、Mapper 等业务逻辑），然后通过 ViewResolver 解析 ModelAndView 对象，并将结果返回给用户。

10.3 Spring MVC 的简单应用

10.1 节和 10.2 节讲解了 Spring MVC 的基本概念，本节将通过实例讲解 Spring MVC 的简单应用，读者需要熟练掌握本节全部内容。

1．创建项目与目录结构

创建基本的 Web 项目，相关操作可参考 2.3.2 小节。目录采用标准的 MVC 结构进行创建，具体如图 10.3 所示。

2．引入 JAR 包

Spring MVC 依赖于 Spring，在此将 Spring、Spring MVC 和 MyBatis 需要的 JAR 包全部引入。需要引入的 JAR 包如图 10.4 所示。

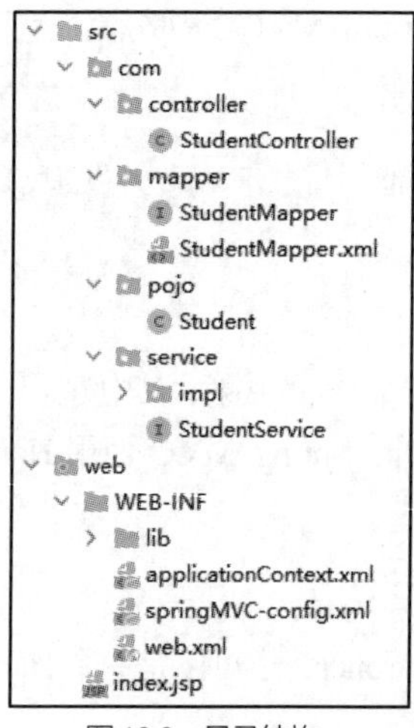

图 10.3 目录结构

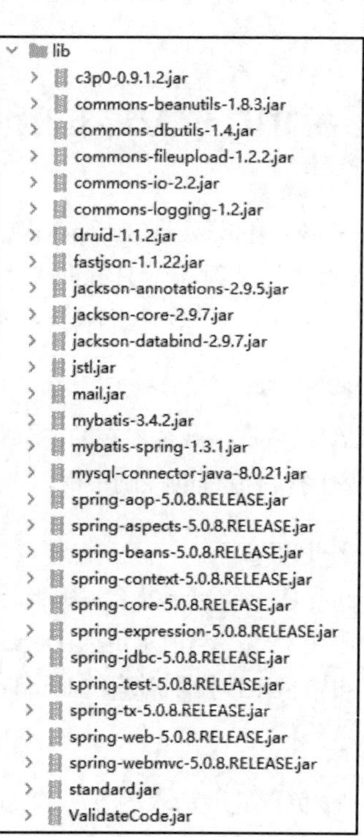

图 10.4 需要引入的依赖包

3．创建 applicationContext.xml 文件

在 WEB-INF 文件夹下创建 applicationContext.xml 文件，在其中添加基本配置，添加的代码如下所示。

```
<!--扫描 com 包下的所有注解-->
<context:component-scan base-package="com"/>
```

4．创建 springMVC-config.xml 文件

在 WEB-INF 文件夹下创建 springMVC-config.xml 文件，其中的代码如例 10-1 所示。

【例 10-1】 springMVC-config.xml

```
1.  <?xml version="1.0" encoding="UTF-8"?>
2.  <beans xmlns="http://www.springframework.org/schema/beans"
3.        xmlns:xsi="http://www.w3.org/2001/XMLSchema-instance"
4.        xmlns:context="http://www.springframework.org/schema/context"
5.        xmlns:mvc="http://www.springframework.org/schema/mvc"
6.        xsi:schemaLocation="http://www.springframework.org/schema/beans
7.      http://www.springframework.org/schema/beans/spring-beans.xsd
8.      http://www.springframework.org/schema/context
9.      http://www.springframework.org/schema/context/spring-context.xsd
10.     http://www.springframework.org/schema/mvc
11.     http://www.springframework.org/schema/mvc/spring-mvc-3.0.xsd">
12.
13.     <!--开启 Spring MVC 注解模式-->---------
14.     <mvc:annotation-driven />
15.     <!—配置静态资源访问-->
16.     <mvc:default-servlet-handler/>
17.     <!--自动扫描相关的包-->
18.     <context:component-scan base-package="com.controller"/>
19.     <!-- 配置视图解析器 -->
20.     <bean class="org.springframework.web.servlet.view.
InternalResourceViewResolver">
21.         <property name="suffix" value=".jsp" />
22.     </bean>
23. </beans>
```

例 10-1 中的文件为 Spring MVC 的配置文件，下面详细介绍其中的内容。

（1）在代码的第 14 行开启 Spring MVC 的注解模式。

（2）在代码的第 16 行开启静态页面的访问。

（3）在代码的第 20～22 行配置视图解析器，视图解析器的作用是在控制器返回的视图名称中自动添加.jsp 后缀。

5．创建 web.xml 文件

在 WEB-INF 文件夹下创建 web.xml 文件，其中的代码如例 10-2 所示。

【例 10-2】 web.xml

```
1.  <!DOCTYPE web-app PUBLIC
2.          "-//Sun Microsystems, Inc.//DTD Web Application 2.3//EN"
3.          "http://java.sun.com/dtd/web-app_2_3.dtd" >
4.  <web-app>
5.      <display-name>Archetype Created Web Application</display-name>
6.      <!--指定 Spring 配置文件的位置-->
7.      <context-param>
8.          <param-name>contextConfigLocation</param-name>
9.          <param-value>WEB-INF/applicationContext.xml</param-value>
10.     </context-param>
```

```
11.     <!--添加监听器-->
12.     <listener>
13.         <listener-class>
14.             org.springframework.web.context.ContextLoaderListener
15.         </listener-class>
16.     </listener>
17.
18.     <!--Spring 核心控制器-->
19.     <servlet>
20.         <servlet-name>dispatcher</servlet-name>
21.         <servlet-class>
22.             org.springframework.web.servlet.DispatcherServlet
23.         </servlet-class>
24.         <init-param>
25.             <param-name>contextConfigLocation</param-name>
26.             <param-value>/WEB-INF/springMVC-config.xml</param-value>
27.         </init-param>
28.         <load-on-startup>1</load-on-startup>
29.     </servlet>
30.
31.     <servlet-mapping>
32.         <servlet-name>dispatcher</servlet-name>
33.         <!--拦截所有的 Mapping，即后台 Controller 程序的所有路径-->
34.         <url-pattern>/</url-pattern>
35.     </servlet-mapping>
36.
37.     <!--    默认访问首页-->
38.     <welcome-file-list>
39.         <welcome-file>index.jsp</welcome-file>
40.     </welcome-file-list>
41. </web-app>
```

例 10-2 中代码的第 7～10 行指定 Spring 配置文件的位置，代码的第 21～23 行配置 DispatcherServlet，代码的第 24～27 行指定 Spring MVC 配置文件的位置。

6．创建 StudentController.java 文件

为了验证 Web 访问是否成功，在 controller 文件夹下创建 StudentController.java 文件，其中的代码如例 10-3 所示。

【例 10-3】 StudentController.java

```
1.  /*标注 StudentController 类为控制器类*/
2.  @Controller
3.  public class StudentController {
4.      /*返回值为 JSON 字符串*/
5.      @ResponseBody
6.      /*访问地址*/
7.      @RequestMapping("/getStu")
8.      public Student getNewStudent(){
9.          /*返回一个 Student 对象*/
10.         return new Student("zhangsan");
11.     }
12. }
```

下面详细讲解例 10-3 中的代码。

（1）在代码的第 2 行使用@Controller 注解标注 StudentController 类为控制器类。

（2）在代码的第 5 行使用@ResponseBody 注解标注 getNewStudent()方法，表示此方法的返回值为 JSON 字符串。如果不加此注解，Spring MVC 会寻找返回值对应的 JSP 文件。

（3）在代码的第 7 行使用@RequestMapping 注解来指定访问地址，当此地址被访问时，将会执行被此注解标注的方法。

7．发布并运行项目

将 war 包发布到 Tomcat 中，运行 Tomcat，在浏览器中输入以下地址。

```
http://localhost:8080/springMVCTest/getStu
```

运行结果如图 10.5 所示。

```
id:      null
name:    "zhangsan"
```

图 10.5　运行结果

从图 10.5 中可以看出，此次请求成功，并以 JSON 字符串的形式返回新建的 Student 对象。

10.4 Spring MVC 的常用注解

为了简化开发，Spring MVC 提供了一系列注解供开发人员使用，除了 10.3 节中讲解的@RequestMapping 和@ResponseBody 注解，Spring MVC 还提供了@RequestParam、@PathVariable、@CookieValue、@RequestBody 和@RequestHeader 等注解。本节将详细介绍 Spring MVC 的常用注解。

10.4.1 @RequestMapping 注解

@RequestMapping 注解用于处理请求地址的映射，它可以用于类，也可以用于方法。当@RequestMapping 注解用于类时，表示地址映射需要将 value 值指定的地址作为父地址；当@RequestMapping 注解用于方法时，表示地址映射需要将 value 值指定的地址作为子地址。@RequestMapping 注解有许多重要的属性，如表 10.1 所示。

表 10.1　@RequestMapping 注解的属性

属性	说明
value	指定请求的地址，当只有 value 属性时，value 属性名可以省略
method	指定该注解标注的方法可以处理的 HTTP 请求方式，如果没有指定 method 属性，则默认映射所有的 HTTP 请求方式
consumes	指定处理请求的提交内容类型（Content-Type）
produces	设置返回值类型和返回值编码
params	指定请求参数中必须包含的参数
headers	设置请求的请求头
name	为映射的地址指定别名

下面通过示例讲解@RequestMapping 注解中常用属性的使用方法。

1．value 属性

修改例 10-3 中的代码，修改后的代码如下所示。

```
1.   @ResponseBody
2.   //此语句等效于@RequestMapping(value="/getNewStudentTestValue")
```

```
3.  @RequestMapping("/getNewStudentTestValue")
4.  public Student getNewStudentTestValue(){
5.      /*返回一个 Student 对象*/
6.      return new Student("zhangsan");
7.  }
```

从以上代码中可以看出，value 属性是@RequestMapping 注解的默认属性。如果只指定了此属性，则可以省略其属性名；如果@RequestMapping 注解指定了多个属性，则必须加上 value 属性名。此外，value 属性名支持通配符匹配，代码如下所示。

```
1.  @ResponseBody
2.  @RequestMapping("/getNewStudentTestValue/*")
3.  public Student getNewStudentTestValue(){
4.      /*返回一个 Student 对象*/
5.      return new Student("zhangsan");
6.  }
```

以上代码使用“*”通配符匹配任意字符，以上代码可以映射以“getNewStudentTestValue”为父地址的所有请求。

运行 Tomcat，在浏览器中访问@RequestMapping 注解指定的地址，运行结果如图 10.6 所示。

图 10.6　@RequestMapping 注解中 value 属性的测试结果

2．method 属性

method 属性用来指定用@RequestMapping 注解标注的方法能够处理的 HTTP 请求方式，示例代码如下所示。

```
1.  @ResponseBody
2.  @RequestMapping(
3.          value = "/getNewStudentTestMethod",
4.          method = {RequestMethod.POST}
5.  )
6.  public Student getNewStudentTestMethod(){
7.      return new Student("zhangsan:Method Test");
8.  }
```

第 4 行代码使用 method 属性来指定 getNewStudentTestMethod()方法只有 POST 请求才可以映射，GET、DELETE 和 PUT 等方式的请求将被拒绝映射。如果没有指定 method 属性，则该方法支持全部的 HTTP 请求方式。

运行 Tomcat，在浏览器中访问@RequestMapping 注解指定的地址，运行结果如图 10.7 所示。

图 10.7　@RequestMapping 注解中 method 属性（POST）的测试结果

从图 10.7 中可以看出，GET 方式不允许访问。将 method 属性中的“POST”改为“GET”再次访问指定的地址，运行结果如图 10.8 所示。

图 10.8 @RequestMapping 注解中 method 属性（GET）的测试结果

从图 10.8 中可以发现，GET 方式请求成功。

3．consumes 属性

consumes 属性用于指定处理请求的提交内容类型（Content-Type），示例代码如下所示。

```
1.  @ResponseBody
2.  @RequestMapping(
3.          value = "/getNewStudentTestConsumes",
4.          consumes = "application/json"
5.  )
6.  public Student getNewStudentTestConsumes(){
7.      return new Student("zhangsan:Consumes Test");
8.  }
```

第 4 行代码使用 consumes 属性来指定请求的 Content-Type，只有对应类型的请求才会被映射。运行 Tomcat，在浏览器中访问@RequestMapping 注解指定的地址，运行结果如图 10.9 所示。

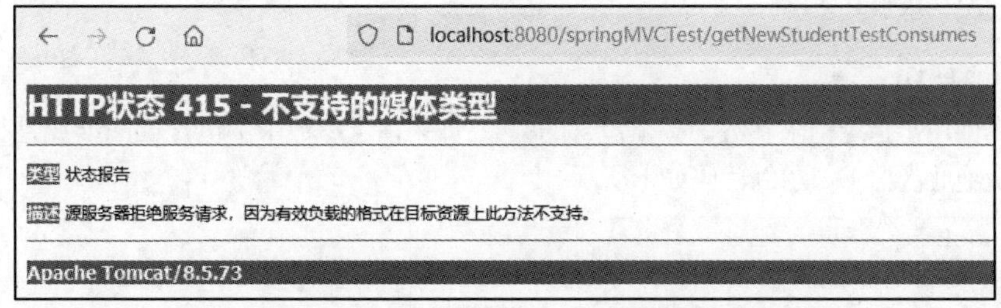

图 10.9 @RequestMapping 注解中 consumes 属性的测试结果

因为 GET 方式请求的请求头中没有 Content-Type，不符合 consumes 属性指定的类型，所以此次请求失败。

4．produces 属性

produces 属性负责指定 response 的 Content-Type。它规定请求返回的内容类型必须是 request 请求头中包含的类型，代码如下所示。

```
1.  @ResponseBody
2.  @RequestMapping(
3.          value = "/getNewStudentTestProduces",
4.          produces = "application/json"
5.  )
6.  public Student getNewStudentTestProduces(){
7.      return new Student("zhangsan:Produces Test");
8.  }
```

第 4 行代码使用 produces 属性来指定 response 的 Content-Type。此外，利用 produces 属性还可以

指定返回值的编码，代码如下所示。

```
1.  @RequestMapping(
2.      value="getNewStudentTestProduces",
3.      produces ="application/json;charset=utf-8"
4.  )
```

以上代码指定返回值的编码是 utf-8，如果不是 utf-8 编码的返回值，服务器将拒绝此次请求。运行 Tomcat，在浏览器中访问@RequestMapping 注解指定的地址，运行结果如图 10.10 所示。

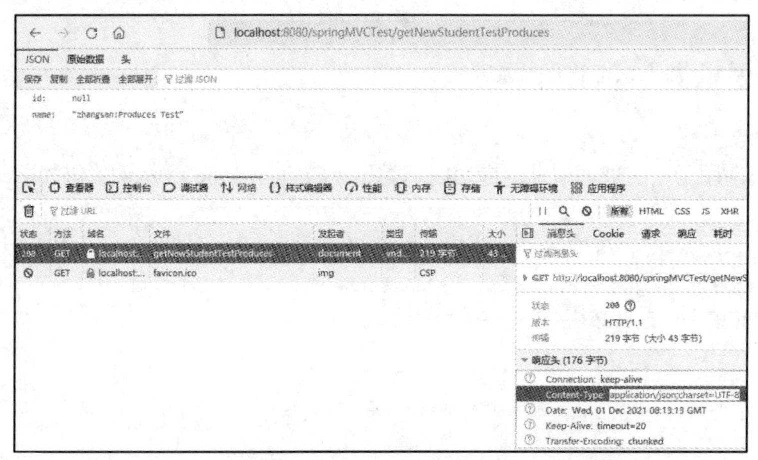

图 10.10 @RequestMapping 中 produces 属性的测试结果

从图 10.10 中可以看出，Content-Type 的值为“application/json;charset=UTF-8”，符合 produces 属性指定的值，此次请求成功。

5．params 属性

params 属性用于指定请求参数中必须包含的参数，如果请求参数中不包含 params 属性指定的参数，则服务器将拒绝此次请求，示例代码如下所示。

```
1.  @ResponseBody
2.  @RequestMapping(
3.          value = "/getNewStudentTestParams",
4.          params = {"id"}
5.  )
6.  public Student getNewStudentTestParams(Integer id){
7.      return new Student(id,"zhangsan:Params Test");
8.  }
```

第 4 行代码使用 params 属性来指定请求参数中必须包含 id 参数。运行 Tomcat，在浏览器中访问@RequestMapping 注解指定的地址，运行结果如图 10.11 所示。

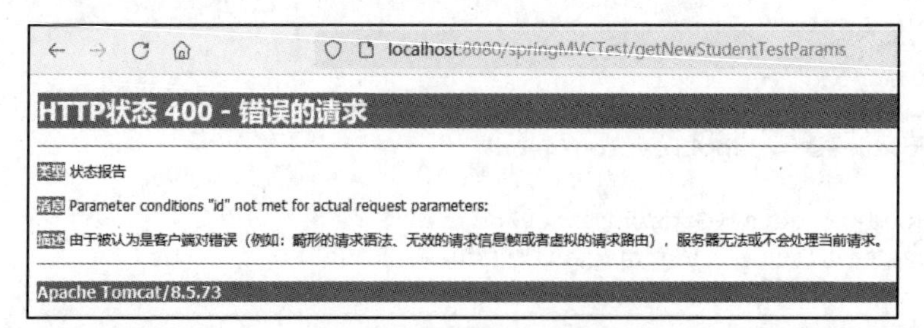

图 10.11 @RequestMapping 注解中 params 属性（请求参数中不包含 id 参数）的测试结果

从图 10.11 中可以看出，当请求参数中不包含 id 参数时，服务器会拒绝此请求。下面将 id 参数添加至请求参数中，再次访问指定的地址，运行结果如图 10.12 所示。

图 10.12 @RequestMapping 注解中 params 属性（请求参数中包含 id 参数）的测试结果

从图 10.12 中可以看出，此次请求成功，并且返回了对应 id 的学生信息。

6．headers 属性

headers 属性的作用是指定请求的请求头，示例代码如下所示。

```
1.  @ResponseBody
2.  @RequestMapping(
3.          value = "/getNewStudentTestHeaders",
4.          headers = {"Host=http://127.0.0.1"}
5.  )
6.  public Student getNewStudentTestHeaders(){
7.      return new Student("zhangsan:Headers Test");
8.  }
```

第 4 行代码使用 headers 属性来指定请求头的 Host 属性。只有请求头中的 Host 属性为 “http://127.0.0.1” 时才会处理该请求。

运行 Tomcat，在浏览器中访问@RequestMapping 注解指定的地址，运行结果如图 10.13 所示。

图 10.13 @RequestMapping 注解中 headers 属性的测试结果（失败）

从图 10.13 中可以看出，服务器拒绝此请求，此时 headers 属性指定的 Host 属性值为 “localhost:8080”。下面，将代码中的 Host 属性值改为 “localhost:8080”，再次访问指定的地址，运行结果如图 10.14 所示。

图 10.14 @RequestMapping 注解中 header 属性的测试结果（成功）

从图 10.14 中可以看出，此次请求成功。

7．name 属性

name 属性仅起到注释作用，用来说明@RequestMapping 注解标注的方法，代码如下所示。

```
1.  @ResponseBody
2.  @RequestMapping(value = "/getStuById",name = "查看学生")
3.  public Student getStudent(){
4.      return new Student("zhangsan");
5.  }
```

10.4.2 @RequestParam 注解

将@RequestParam 注解添加在方法的参数前，可将请求参数绑定到控制器的方法参数上。这样可以针对参数进行限制与赋值操作。修改例 10-3 的代码，修改完成后代码如下所示。

```
1.  @ResponseBody
2.  @RequestMapping(value = "/getStuByIdDefault")
3.  public Student getStudentByIdDefault(
4.          @RequestParam(required = false , defaultValue = "0") Integer id
5.  ){
6.      /*返回一个 Student 对象*/
7.      return new Student(id,"zhangsan");
8.  }
```

第 4 行代码使用@RequestParam 注解限制 id 参数。在@RequestParam 注解中，required 用于设置此参数是否必须传入，默认为 true；defaultValue 用于设置此参数的默认值，当请求中没有此参数的值时，默认使用 defaultValue 指定的值。

运行 Tomcat，在浏览器中访问@RequestMapping 注解指定的地址，运行结果如图 10.15 所示。

图 10.15 @RequestParam 注解无参数的测试结果

从图 10.15 中可以看出，当没有传入 id 参数值时，使用默认 id 参数值，返回 id 为 0 的学生的信息，下面传入 id 参数值，再次访问指定的地址，运行结果如图 10.16 所示。

图 10.16 @RequestParam 注解有参数的测试结果

从图 10.16 中可以看出，当传入 id 参数值时，返回对应 id 的学生的信息。

10.4.3 @PathVariable 注解

@PathVariable 注解是 Spring3.0 中新增的一个注解，它负责接收请求路径中的占位符，修改例 10-3 中的代码，修改后的代码如下所示。

```
1.  @ResponseBody
2.  @RequestMapping(value = "/getStudentByName/{name}")
3.  public Student getStudentByName(
4.          @PathVariable("name") String name
5.  ){
6.      /*返回一个 Student 对象*/
7.      return new Student(name);
8.  }
```

第 2 行代码使用“{name}”预先接收一个参数，然后此参数通过@PathVariable 注解被第 4 行代码的 name 属性接收。当使用“{}”来接收参数时，请求路径中不需要附加属性名，在访问时直接添加需要传递的值即可。

运行 Tomcat，在浏览器中访问@RequestMapping 注解指定的地址，运行结果如图 10.17 所示。

图 10.17 @PathVariable 注解的测试结果

从图 10.17 中可以看出，请求路径中没有使用“?name=lisi”来传参，而是直接在路径后添加相应的参数值，此值会被@PathVariable 注解取出并赋给被此注解标注的属性。

10.4.4 @RequestBody 注解

@RequestBody 注解用来处理不以“application/x-www-form-urlcoded”格式编码的提示内容，“application/json”格式就是其中的一种。在实际开发中，POST 与 PUT 请求通常以“application/json”格式传递数据，此时就需要使用@RequestBody 注解来接收数据。下面将以 POST 请求为例，编写接收数据的代码，代码如下所示。

```
1.  @ResponseBody
2.  @RequestMapping(value = "/addStudent")
3.  public Student addStudent(
4.      @RequestBody Student student
5.  ){
6.      return student;
7.  }
```

在以上代码中，使用@RequestBody 注解将接收到的请求体中的内容封装为 Student 对象。使用请求工具发送 POST 请求（此部分的知识不需要读者掌握，仅了解即可），运行结果如图 10.18 所示。

从图 10.18 中可以看出，发送 POST 请求后，@RequestBody 注解将发送的数据全部接收，并将其封装为对象，然后将封装好的 Student 对象返回给前端。

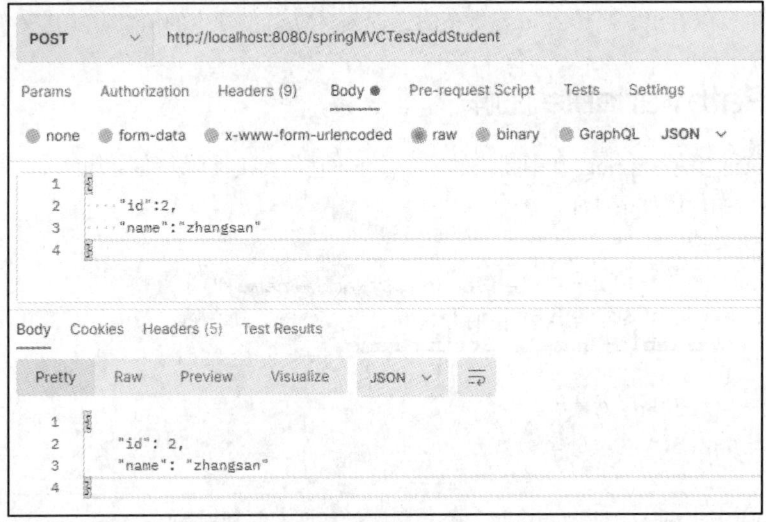

图 10.18 @RequestBody 注解测试结果

10.5 Spring MVC 中的参数绑定

在 Spring MVC 中，提交请求的数据是通过方法的形参来接收的。例如，在 10.4.3 小节中，使用@PathVariable 注解可以接收基本类型或者 String 类型的数据；在 10.4.4 小节中，使用@RequestBody 注解可以接收封装类型的数据。除此之外，Spring MVC 还支持多种数据绑定。本节将详细介绍 Spring MVC 中常用的参数绑定类型，读者需要熟练掌握本节全部内容。

10.5.1 默认数据类型的数据绑定

Servlet 环境有 4 个常用的对象，分别为 HttpServletRequest 对象、HttpServletResponse 对象、HttpSession 对象和 Model 对象。它们被称为 Spring MVC 的默认数据类型，当需要对这些对象进行操作时，可以通过方法参数直接取用，Spring MVC 会自动将相应的对象注入。下面通过一个示例详细介绍这些对象的绑定方法。

修改例 10-3 中的代码，修改后的代码如下所示。

```
1.  @ResponseBody
2.  @RequestMapping(value = "/testDefaultType")
3.  public Map testDefaultType(
4.          HttpServletRequest request,
5.          HttpServletResponse response,
6.          HttpSession session
7.  ){
8.      String host = request.getHeader("Host");
9.      response.setHeader("testDefaultType","test");
10.     session.setAttribute("testDefaultType","test");
11.
12.     Map map = new HashMap();
13.     map.put("主机地址",host);
14.     return map;
15. }
```

在以上代码中，默认数据类型的接收不需要前端传递相应的参数，Spring MVC 会自动将其填充。第 8 行代码的作用为从 request 中获取主机地址，第 9 行代码的作用为设置 response 的 Header 值，第 10 行代码的作用为设置客户端的 session。

运行 Tomcat，在浏览器中访问@RequestMapping 注解指定的地址，运行结果如图 10.19 所示。

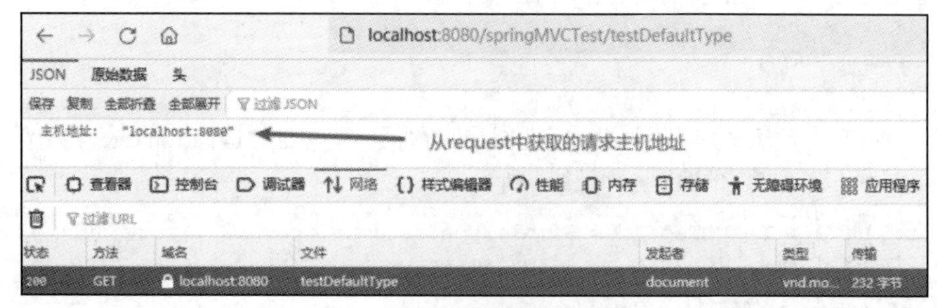

图 10.19 默认数据类型的测试结果

从图 10.19 中可以看出，request 和 response 的操作得到了验证。下面查看 Cookie，结果如图 10.20 所示。

图 10.20 默认数据类型的 session 测试结果

从图 10.20 中可以看出，session 设置生效（session 验证在此不详细讲解），默认数据类型验证完毕。

10.5.2 简单数据类型的数据绑定

相对于默认数据类型的数据绑定，简单数据类型是需要前端传输、后台接收的。它包括 Java 中的 8 种基本类型与 String 类型。下面通过一个示例详细介绍简单数据类型的数据绑定方法。

修改例 10-3 中的代码，修改后的代码如下所示。

```
1.  /*测试简单数据类型的数据绑定*/
2.  @ResponseBody
3.  @RequestMapping(value = "/testSimpleType")
4.  public String testSimpleType(
5.      byte param1,
6.      short param2,
7.      int param3,
8.      long param4,
9.      float param5,
10.     double param6,
11.     char param7,
12.     boolean param8,
13.     String param9
14. ){
15.
16.     System.out.println("byte "+param1);
17.     System.out.println("short "+param2);
18.     System.out.println("int "+param3);
```

```
19.     System.out.println("long "+param4);
20.     System.out.println("float "+param5);
21.     System.out.println("double "+param6);
22.     System.out.println("char "+param7);
23.     System.out.println("boolean "+param8);
24.     System.out.println("String "+param9);
25.
26.     return "ok";
27. }
```

以上代码使用 8 种基本类型和 String 类型来接收前端传来的参数，运行 Tomcat，在浏览器中访问@RequestMapping 注解指定的地址，地址如下所示。

```
localhost:8080/springMVCTset/testSimpleType?param1=1&param2=1&param3=1&param4=
134242342343&param5=1.4&param6=1.44&param7=t&param8=true&param9=string
```

输出结果如图 10.21 所示。

```
byte 1
short 1
int 1
long 134242342343
float 1.4
double 1.44
char t
boolean true
String string
```

图 10.21 输出结果

从图 10.21 中可以看出，所有简单数据类型都可以接收。需要注意的是，基本类型的参数不可以设置为空值，如果前端传递的参数中没有对应的值，则此次请求会失败。

10.5.3 实体 Bean 类型的数据绑定

除了简单数据类型外，Spring MVC 还可以接收实体类型参数的传输，其中，对象属性的名称需要与参数名称一一对应。

实体 Bean 类型的数据绑定方法请参考 10.4.4 小节的示例。

10.5.4 集合数组类型的数据绑定

当需要集合类型参数的传输时，使用@RequestParam 注解指定集合中参数的名称，修改例 10-3 中的代码，修改后的代码如下所示。

```
1.  @ResponseBody
2.  @RequestMapping(value = "/testListArray")
3.  public Map testListArray(
4.          @RequestParam("list") List<String> list,
5.          @RequestParam("strings") String[] strings
6.  ){
7.      Map map = new HashMap();
8.      map.put("list",list);
9.      map.put("strings",strings);
10.
11.     return map;
12. }
```

第 4 行代码使用 List 集合来接收同名参数，使用@RequestParam 注解标注 List 集合中元素的名称为 list。第 5 行代码使用 String 数组来接收同名参数，使用@RequestParam 注解标注 String 数组中元素的名称为 strings。运行 Tomcat，在浏览器中访问@RequestMapping 注解指定的地址，运行结果如图 10.22 所示。

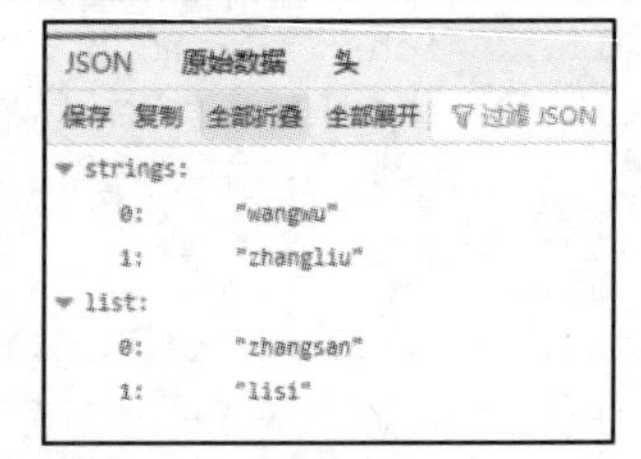

图 10.22 集合数组类型测试结果

从图 10.22 中可以看出，访问成功。除此之外，还有复杂类型的映射，不过，在实际开发中，复杂类型的映射一般以 JSON 字符串形式传递参数，不使用 Spring MVC 的自动映射。

10.6 Spring MVC 中复杂类型的传输

10.5 节讲解了常见类型的数据绑定，当需要传输的数据类型比较复杂时，使用数据绑定就非常麻烦。本节将介绍使用 JSON 字符串来传输复杂数据的方法，读者需要重点掌握本节内容。

使用 10.3 节中的示例，修改例 10-3 中的代码，修改后的代码如下所示。

```
1.  @ResponseBody
2.  @RequestMapping(value = "/testJSONFormat")
3.  public List<Student> testJSONFormat(@RequestBody Map map){
4.      /*创建一个学生集合*/
5.      List<Student> myStudentList = new LinkedList<>();
6.      /*解析 JSON*/
7.      List<Map> studentList = (List) map.get("studentList");
8.
9.      for (Map map1 : studentList) {
10.         int id = (int) map1.get("id");
11.         String name = (String) map1.get("name");
12.         /*向学生集合中添加学生信息*/
13.         System.out.println("向集合中添加学生姓名: "+name);
14.         myStudentList.add(new Student(id,name));
15.
16.     }
17.     return myStudentList;
18. }
```

第 3 行代码使用@RequestBody 注解来接收 POST 类型的请求数据，将接收到的数据封装成 Map 集合，其中，JSON 字符串的键会被解析为 Java 中 Map 集合的键，JOSN 字符串的值会被解析为 Map 集合的值，JSON 字符串的数组会被解析成 List 集合。

使用工具 Postman（不要求掌握）来模拟 POST 请求，结果如图 10.23 所示。

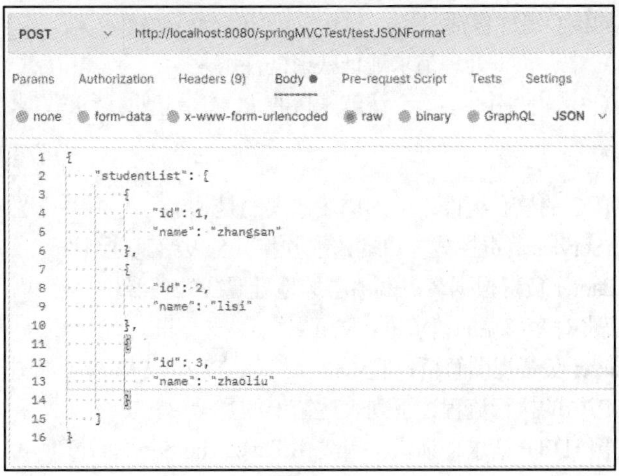

图 10.23　复杂类型的 POST 请求

从图 10.23 中可以看出，发送的 POST 请求携带了一个对象数组，此数组名为 studentList。此数组中有 3 个对象，每个对象都包含 id 和 name 键值对。当此请求传输到后台后，控制台的输出结果如图 10.24 所示。

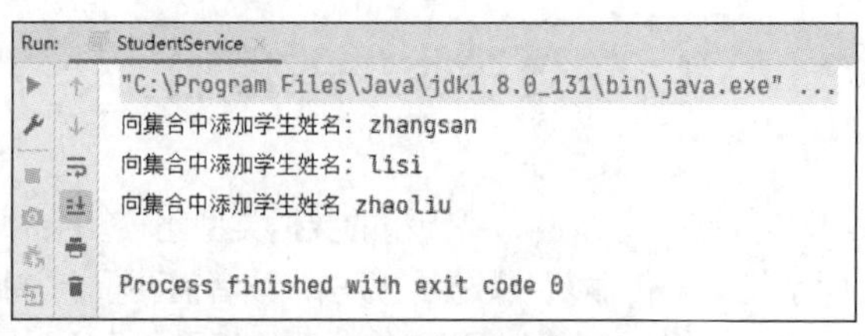

图 10.24　复杂类型的输出结果

从图 10.24 中可以看出，对象数组的数据接收完毕。在日常开发中，使用 JSON 字符串来传递数据的方法经常使用，这种方法需要通过前后台沟通，确定数据格式，根据确定的数据格式编写对应的数据接收方法。

10.7 本章小结

本章首先介绍了 Spring MVC 的基本概念和核心组件，然后通过一个示例讲解了 Spring MVC 的基本使用方法，接着讲解了 Spring MVC 的常用注解和参数绑定方法，最后讲解了开发中常用的 JSON 传输方式。除 Spring MVC 概述外，本章其余部分均需要读者重点掌握。

在本章的内容中，Spring MVC 概述仅需要读者初步理解；Spring MVC 的注解、参数绑定和复杂类型传输是开发中常用的基础知识，需要读者熟练掌握；Spring MVC 的核心组件是面试中常考的知识点，感兴趣的读者可以深入探究每个组件的执行流程。

10.8 习题

1．填空题

（1）Spring MVC 的核心组件包括________、________、________、________和________。

（2）在 Spring MVC 中，负责处理请求地址映射的注解是________。

（3）在 Spring MVC 中，使用________注解来接收 POST 类型的请求数据。

2．选择题

（1）关于 Spring MVC 的核心组件，下列描述错误的是（　　）。

A. DispatcherServlet 负责拦截客户端请求并将其分发给其他组件

B. HandlerAdapter 负责根据客户端请求的地址寻找处理器

C. Handler 负责对客户端的请求进行处理

D. ViewResolver 负责视图解析，它可以将处理结果生成 View（视图）

（2）关于 Spring MVC 的执行流程，下列表述错误的是（　　）。

A. 客户端发出 HTTP 请求，请求将首先由 DispatcherServlet 处理

B. DispatcherServlet 接收到请求后，通过对应的 HandlerAdapter 解析出目标 Handler

C. DispatcherServlet 通过 ViewResolver 完成逻辑视图到真实 View 对象之间的解析

D. DispatcherServlet 将最终的 View 对象响应给客户端并展示给用户

（3）在@RequestMapping 注解的属性中，用于指定该注解标注的方法可以处理的 HTTP 请求方式的是（　　）。

A. value　　B. method　　C. consumes　　D. params

（4）（　　）注解用于将控制器类的方法返回的对象转换为 JSON 响应客户端。

A. @ResponseBody　　B. @RequestBody

C. @RequestMapping　　D. @Controller

（5）（　　）注解可以解析请求的 JSON 格式的参数，以解析 JSON 字符串。

A. @ResponseBody　　B. @RequestBody

C. @RequestMapping　　D. @Controller

3．思考题

（1）简述 Spring MVC 的执行流程。

（2）简述当请求方式为 POST 时，接收参数的方式。

4．编程题

创建 Dog 类，在其中添加 name 和 age 属性。使用 Spring MVC 接收如下请求的参数并将其封装为 Dog 对象。

```
http://localhost:8080/testDog?name=wangcan&age=3 //GET 方式
http://localhost:8080/testDog/wangcai/3 //GET 方式
http://localhost:8080/testDog?myName=wangcan&myAge=3 //GET 方式
http://localhost:8080/testDog //POST 方式，请求体中存在 name 和 age 参数
```

第 11 章 Spring MVC 进阶

本章学习目标

- 了解 Spring MVC 的文件上传和下载功能。
- 了解 Spring MVC 的拦截器。
- 掌握 Spring MVC 实现 RESTful 风格的方法。
- 掌握全局异常的处理方法。

Spring MVC 进阶

第 10 章主要介绍了 Spring MVC 的基础知识，本章将对 Spring MVC 的进阶知识点详细介绍。本章的主要内容包括文件上传与下载、拦截器、RESTful 风格和全局异常处理，其中，RESTful 风格需要读者重点掌握。

11.1 文件上传与下载

文件上传与下载功能的实现需要依赖包的支持，因此要先导入对应的依赖包，需要导入的包如下所示。

```
commons-fileupload-1.3.1.jar
commons-io-2.4.jar
```

采用 10.3 节中的示例，在 controller 文件夹下新建 TestFileUploadAndDownload 类，其代码如例 11-1 所示。

【例 11-1】 TestFileUploadAndDownload.java

```
1.   package com.controller;
2.
3.   @Controller
4.   public class TestFileUploadAndDownload {
5.
6.      /*文件上传*/
7.      @RequestMapping("fileUpload")
8.       public void fileUpload(
9.              @RequestParam("multipartFile") MultipartFile multipartFile,
10.             HttpServletResponse httpServletResponse
11.     ) throws IOException {
12.
13.        String originalFilename = multipartFile.getOriginalFilename();
14.        /*创建一个空文件*/
15.        File file = new File("D://testImage",originalFilename);
16.
17.        /*如果不存在文件，则创建一个文件*/
```

```
18.         if (!file.exists()){
19.            file.mkdirs();
20.         }
21.         file.createNewFile();
22.         /*将接收到的文件输出到指定位置*/
23.         multipartFile.transferTo(file);
24.
25.         httpServletResponse.getWriter().println("upload OK!");
26.     }
27.     /*文件下载*/
28.     @RequestMapping("fileDownload")
29.     public void fileDownload(
30.           HttpServletResponse httpServletResponse
31.     ) throws IOException {
32.        File file = new File("D://testImage//1.jpg");
33.
34.        //告知浏览器下载文件，而不是直接打开文件，浏览器默认打开文件
35.        httpServletResponse.setContentType("application/x-download");
36.        //设置下载文件的名称
37.        httpServletResponse.addHeader(
38.              "Content-Disposition" ,
39.              "attachment;filename=1.jpg"
40.        );
41.        /*使用 FileUtils 类的 readFileToByteArray()方法将 file 文件转换为字节数组*/
42.        byte[] bytes = FileUtils.readFileToByteArray(file);
43.        /*将信息返回*/
44.        httpServletResponse.getOutputStream().write(bytes);
45.     }
46. }
```

下面详细讲解例 11-1 中的代码。

1. 文件上传

（1）代码的第 9 行使用 MultipartFile 类来接收上传的文件，请求上传文件的类型为 POST 方式，可以在 RequestMapping 注解中加入“method = RequestMethod.POST”来限定请求的方式为 POST，此处不附加此代码表示默认接收全部类型的请求。

（2）当控制器接收到文件时，代码的第 13 行通过 getOriginalFilename()方法获取上传文件的名称，并将其作为存放文件的名称。

（3）代码的第 15 行用于创建空文件，此例中将文件存放到 D 盘的 testImage 文件夹下。代码的第 23 行通过 MultipartFile 类的 transferTo()方法将文件输出到指定位置。

2. 文件下载

（1）下载文件通常需要先接收需要下载的文件名称，然后通过文件服务器寻找对应的文件，在此省略寻找文件的过程，直接通过第 32 行的代码定位到对应文件，并将其发送给客户端。

（2）代码的第 35 行用于设置 ContentType 为 application/x-download 类型，此类型是一个下载类型的标记，浏览器在接收到此类型的请求后，直接下载文件。

（3）代码的第 37～40 行设置下载文件的名字。

（4）代码的第 42 行使用 FileUtils 类将文件转换成字节数组。

（5）代码的第 44 行获取 httpServletResponse 对象，调用输出流，将转换得到的字节数组输出到前端。

3．创建 index.jsp 文件

在 Web 文件夹下创建 index.jsp 文件，编写其中的代码，代码如下所示。

```
1.  <%@ page contentType="text/html;charset=UTF-8" language="java" %>
2.  <html>
3.    <head>
4.    </head>
5.    <body>
6.      <form action="/fileUpload" method="post" enctype="multipart/form-data">
7.        <input type="file" name="multipartFile">
8.        <input type="submit" value="确定">
9.      </form>
10.   </body>
11. </html>
```

以上代码的第 7 行用于指定文件的名称，此名称需要与例 11-1 的第 9 行代码中@RequestParam 注解指定的名称相同。

4．测试文件上传功能

发布项目，运行 Tomcat，打开浏览器访问项目地址，上传文件页面如图 11.1 所示。

选择图片文件 1.jpg，单击“确定”按钮上传图片。观察 D 盘中 testImage 文件夹下的图片，如图 11.2 所示。

图 11.1　上传文件页面

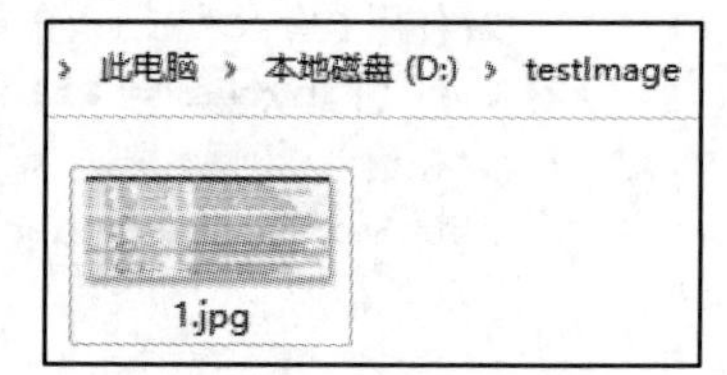

图 11.2　上传的文件

从图 11.2 中可以发现，文件 1.jpg 成功传输到指定目录，文件上传功能测试成功。

5．测试文件下载功能

文件下载通过访问相应的下载地址就可以实现。打开浏览器，访问文件下载地址，地址如下所示。

```
http://localhost:8080/fileDownload
```

然后浏览器就可以下载对应的文件，文件下载页面如图 11.3 所示。

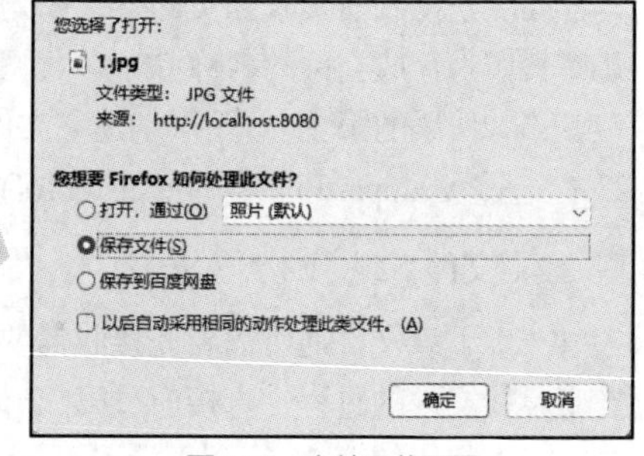

图 11.3　文件下载页面

11.2 拦截器

在 Java Web 中，Filter（过滤器）负责过滤请求。在 Spring MVC 中，除了过滤器之外，还可以使用 Interceptor（拦截器）过滤请求。本节将详细介绍拦截器的使用方法，读者只需了解本节内容即可。

11.2.1 拦截器与过滤器的区别

拦截器与过滤器都可以实现过滤请求的效果，下面介绍它们的不同点。

拦截器是 Spring MVC 提供的，是基于反射生成的，通过函数调用实现其功能；而过滤器是直接创建完成的，是基于函数回调实现其功能的。

同时，在拦截器中可以获取 Spring 容器中的 Bean 对象；而过滤器不依赖 Spring 容器，它依赖的是 Servlet 环境。

11.2.2 拦截器

下面使用一个示例讲解拦截器的使用方法。在此采用 10.3 节示例的环境，在 com 包下新建 config 文件夹，在 config 文件夹下新建 MyInterceptor 类，使 MyInterceptor 类实现 HandlerInterceptor 接口，使用 Alt+Ins 组合键生成需要覆盖的方法，代码如例 11-2 所示。

【例 11-2】 MyInterceptor.java

```
1.  public class MyInterceptor implements HandlerInterceptor {
2.      @Override
3.      public boolean preHandle(
4.              HttpServletRequest request,
5.              HttpServletResponse response,
6.              Object handler
7.      ) throws Exception {
8.          return true;
9.      }
10.
11.     @Override
12.     public void postHandle(
13.             HttpServletRequest request,
14.             HttpServletResponse response,
15.             Object handler,
16.             ModelAndView modelAndView
17.     ) throws Exception {
18.
19.     }
20.
21.     @Override
22.     public void afterCompletion(
23.             HttpServletRequest request,
24.             HttpServletResponse response,
25.             Object handler,
26.             Exception ex
27.     ) throws Exception {
28.
29.     }
30. }
```

下面详细讲解例 11-2 中的代码。

（1）preHandle()方法在目标方法执行前执行，可以在业务代码执行之前控制请求。

（2）postHandle()方法在目标方法执行完毕后、视图返回前执行，通过此方法可以获取 ModelAndView 视图并进行修改，同时，也可以修改 request 和 response。

（3）afterCompletion()方法在视图渲染后执行，通过此方法可以控制最终返回给客户端的信息。

需要注意的是，拦截器可以有多个，但其中只要有一个拦截器的 preHandle()方法返回 false，则此请求不再向后执行。

拦截器创建完成后，需要在 Spring MVC 配置文件中注册此拦截器，代码如下所示。

```
1.  <mvc:interceptors>
2.      <!-- 配置拦截器 -->
3.      <mvc:interceptor>
4.          <mvc:mapping path="/**"/>
5.          <!-- 配置拦截器对象 -->
6.          <bean class="com.config.MyInterceptor"/>
7.      </mvc:interceptor>
8.  </mvc:interceptors>
```

在以上代码中，通过<mvc:interceptors>元素配置拦截器集。在拦截器集中，可以配置多个具体的拦截器，在<mvc:interceptor>元素中可以配置拦截器的地址和拦截器拦截的地址。

11.2.3 拦截器的执行流程

为了更加清晰地讲解拦截器在 Web 环境中的执行流程，在下面的示例中加入过滤器，通过一个完整的示例讲解拦截器的执行流程。

1．初始化环境

创建 Web 应用目录，导入 10.3 节引入的 JAR 包。

2．创建 MyFilter 类

在 com 文件夹下创建 config 文件夹，在 config 文件夹下创建 MyFilter 类，代码如例 11-3 所示。

【例 11-3】 MyFilter.java

```
1.  public class MyFilter implements Filter {
2.      @Override
3.      public void init(FilterConfig filterConfig) throws ServletException {
4.
5.      }
6.
7.      @Override
8.      public void doFilter(
9.              ServletRequest servletRequest,
10.             ServletResponse servletResponse,
11.             FilterChain filterChain
12.     ) throws IOException, ServletException {
13.         System.out.println("过滤器 在方法前执行");
14.         filterChain.doFilter(servletRequest,servletResponse);
15.         System.out.println("过滤器 在方法后执行");
16.     }
17.
18.     @Override
19.     public void destroy() {
20.
21.     }
22. }
```

在以上代码中，通过过滤器在方法执行前后输出相应的语句，然后在 web.xml 文件中注册此过滤器，代码如下所示。

```
1.  <filter>
2.      <filter-name>MyFilter</filter-name>
3.      <filter-class>com.config.MyFilter</filter-class>
```

```
4.  </filter>
5.  <filter-mapping>
6.      <filter-name>MyFilter</filter-name>
7.      <url-pattern>/*</url-pattern>
8.  </filter-mapping>
```

3. 创建拦截器类

在 config 文件夹下创建两个拦截器类，代码分别如例 11-4 和例 11-5 所示。

【例 11-4】 MyInterceptor.java

```
1.  public class MyInterceptor implements HandlerInterceptor {
2.      @Override
3.      public boolean preHandle(
4.              HttpServletRequest request,
5.              HttpServletResponse response,
6.              Object handler
7.      ) throws Exception {
8.          System.out.println("first preHandle 执行");
9.          return true;
10.     }
11.
12.     @Override
13.     public void postHandle(
14.             HttpServletRequest request,
15.             HttpServletResponse response,
16.             Object handler,
17.             ModelAndView modelAndView
18.     ) throws Exception {
19.         System.out.println("first postHandle 执行");
20.     }
21.
22.     @Override
23.     public void afterCompletion(
24.             HttpServletRequest request,
25.             HttpServletResponse response,
26.             Object handler,
27.             Exception ex
28.     ) throws Exception {
29.         System.out.println("first afterCompletion 执行");
30.     }
31. }
```

【例 11-5】 MySecondInterceptor.java

```
1.  public class MySecondInterceptor implements HandlerInterceptor {
2.      @Override
3.      public boolean preHandle(
4.              HttpServletRequest request,
5.              HttpServletResponse response,
6.              Object handler
7.      ) throws Exception {
8.          System.out.println("second preHandle 执行");
9.          return true;
10.     }
```

```
11.
12.     @Override
13.     public void postHandle(
14.            HttpServletRequest request,
15.            HttpServletResponse response,
16.            Object handler,
17.            ModelAndView modelAndView
18.     ) throws Exception {
19.        System.out.println("second postHandle 执行");
20.     }
21.
22.     @Override
23.     public void afterCompletion(
24.            HttpServletRequest request,
25.            HttpServletResponse response,
26.            Object handler,
27.            Exception ex
28.     ) throws Exception {
29.        System.out.println("second afterCompletion 执行");
30.     }
31. }
```

利用例 11-4 与例 11-5 中的 perHandle()、postHandle()和 afterCompletion()方法输出相应的语句，测试执行顺序。

在 Spring MVC 配置文件中添加拦截器，代码如下所示。

```
1.  <mvc:interceptors>
2.     <!-- 配置拦截器 -->
3.     <mvc:interceptor>
4.        <mvc:mapping path="/**"/>
5.        <!-- 配置拦截器对象 -->
6.        <bean class="com.config.MyInterceptor"/>
7.     </mvc:interceptor>
8.     <mvc:interceptor>
9.        <mvc:mapping path="/**"/>
10.       <!-- 配置拦截器对象 -->
11.       <bean class="com.config.MySecondInterceptor"/>
12.    </mvc:interceptor>
13. </mvc:interceptors>
```

4．创建 TestInterceptor 类

在 com 文件夹下的 controller 文件夹中创建 TestInterceptor 类，代码如例 11-6 所示。

【例 11-6】 TestInterceptor.java

```
1.  @Controller
2.  public class TestInterceptor {
3.     @RequestMapping("testInterceptor")
4.     public String test(){
5.        System.out.println("方法执行");
6.        return "index";
7.     }
8.  }
```

在例 11-6 的代码中，未使用@ResponseBody 注解，返回至 index.jsp 对应的页面。

发布项目，启动 Tomcat，访问例 11-6 中的控制器，输出结果如图 11.4 所示。

```
过滤器 在方法前执行
first preHandle 执行
second preHandle 执行
方法执行
second postHandle 执行
first postHandle 执行
second afterCompletion 执行
first afterCompletion 执行
过滤器 在方法后执行
```

图 11.4 拦截器执行流程

从图 11.4 中可以看出，当请求发送到服务器后，首先执行 Filter 中 doFilter()方法之前的代码，然后执行拦截器的 perHandle()z 方法，再执行目标方法。当目标方法执行完毕后，执行 postHandle()方法，然后执行 afterCompletion()方法，最后执行 Filter 中 doFilter()方法之后的代码。

11.3 RESTful 风格

RESTful 是开发中常用的编程风格，它针对每种不同的 URL 类型指定了相应的请求方式。本节将详细介绍 RESTful 风格，读者需要熟练掌握本节全部内容。

11.3.1 RESTful 风格简介

RESTful 风格主要应用于普通的增删改查操作，下面列出了使用 RESTful 风格时需要遵守的规范。

（1）在通过 URL 传递参数时尽量使用参数值直接传参，同时使用@PathVariable 注解来适配属性，避免使用类似“?name=zhangsan”的拼接方式来传递参数。

（2）增删改查操作使用的父请求地址需保持一致。

（3）对于同一个对象的增删改查操作，需要通过 method 属性指定相应的请求类型，每个请求根据不同的请求类型访问对应的方法，具体的对应关系如表 11.1 所示。

表 11.1 RESTful 风格请求类型对应的方法

请求类型	方法
新增	POST
修改	PUT
查询	GET
删除	DELETE

11.3.2 RESTful 风格的实现

下面通过示例来讲解 RESTful 风格的实现。

使用 10.3 节示例的环境，在 controller 文件夹下创建 TestRESRful.java 文件，其代码如例 11-7 所示。

【例 11-7】 TestRESRful.java

```
1.  /*此示例针对 Student 类进行操作*/
2.  /*可以把@RestController 理解为升级版的@Controller,
3.  表示为此 Controller 下的所有方法都加上@ResponseBody 注解*/
4.  @RestController
5.  @RequestMapping("student")
6.  public class TestRESRful {
7.      //此方法的作用相当于@RequestMapping(method= {RequestMethod.POST})的作用
```

```
8.      @PostMapping(produces = {"application/json;charset=utf-8"})
9.      public String addStudent(){
10.         return "添加成功";
11.     }
12.     @GetMapping
13.     public Student getStudent(){
14.         return new Student(1,"zhangsan");
15.     }
16.     @GetMapping (value = "/{id}")
17.     public Student getStudentById(@PathVariable("id") Integer id){
18.         return new Student(id,"zhangsan");
19.     }
20.     @PutMapping (
21.          value = "/{id}",
22.          produces = {"application/json;charset=utf-8"}
23.     )
24.     public String updateStudent(
25.             @PathVariable("id") Integer id,
26.             @RequestBody Map map
27.     ){
28.         String name = (String) map.get("name");
29.         System.out.println("将"+id+"的 name 更改为"+name);
30.         return "更改成功";
31.     }
32.     @DeleteMapping (
33.             value = "/{id}",
34.             produces = {"application/json;charset=utf-8"}
35.     )
36.     public String DeleteStudent(@PathVariable("id") Integer id){
37.         return "删除成功";
38.     }
39. }
```

下面详细介绍例 11-7 中的代码。

（1）代码的第 4 行使用@RestController 注解，标注 TestRESRful 类为控制器，且默认为此类中的所有方法都加上@ResponseBody 注解。

（2）代码的第 8 行，使用@PostMapping 注解标注一个方法，这表示此方法必须是 POST 请求，同理，使用@GetMapping 注解表示此次访问必须使用 GET 方式。此外，如果将@RequestMapping 注解的 value 属性设置为空值，则直接使用父目录（类上方@RequestMapping 注解指定的地址）作为访问地址。

（3）代码的第 22 行使用“{}”来传参，在方法中使用@PathVariable 注解来指定参数。

发布项目，运行 Tomcat，观察日志，如图 11.5 所示。

```
Mapped "{[/student],methods=[POST]}" onto public java.lang.String com.controller.TestRESRful.addStudent()
Mapped "{[/student/{id}],methods=[DELETE]}" onto public java.lang.String com.controller.TestRESRful.DeleteStudent(java.lang.Integer)
Mapped "{[/student/{id}],methods=[PUT]}" onto public java.lang.String com.controller.TestRESRful.updateStudent(java.lang.Integer,java.util.Map)
Mapped "{[/student],methods=[GET]}" onto public com.pojo.Student com.controller.TestRESRful.getStudent()
Mapped "{[/student/{id}],methods=[GET]}" onto public com.pojo.Student com.controller.TestRESRful.getStudentById(java.lang.Integer)
```

图 11.5　Tomcat 发布日志

从图 11.5 中可以看出，对应映射关系已经配置完毕，下面访问对应请求，在此以访问 updateStudent()方法为例，结果如图 11.6 所示。

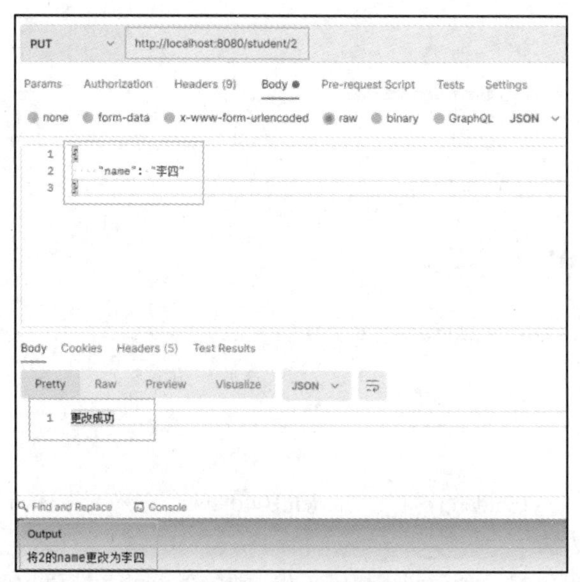

图 11.6 测试更新方法

在图 11.6 中，使用 PUT 请求访问 updateStudent()方法，参数为 name，参数值为“李四”，单击发送按钮，前端返回“更改成功”，后台输出修改信息。

11.4 全局异常处理

异常在开发中经常发生。发布项目时，如果某次请求出现异常，就需要根据异常的类型返回一个合理的结果，此时，针对每个请求都设置异常处理显得极为不妥。Spring MVC 提供了一系列全局异常处理方法。本节将讲解开发中常用的全局异常处理方案，读者需要熟练掌握本节全部内容。

11.4.1 异常处理关键注解

在进行实战之前，读者需要了解两个注解，分别是@ControllerAdvice 注解和@ExceptionHandler 注解。

@ControllerAdvice 注解标注在类上方，表示此类是异常处理类。

@ExceptionHandler 注解标注在异常处理类中的方法上，表示此方法为异常处理方法。在此注解中，需要添加对应的异常类型。此后，只要程序中出现相应的异常，此方法将会执行，返回的结果将直接发送到前台。

11.4.2 全局异常处理示例

本例使用 10.3 节的环境，在 controller 文件夹下创建 TestUnifyException.java 文件来测试异常，代码如例 11-8 所示。

【例 11-8】 TestUnifyException.java

```
1.  @RestController
2.  @RequestMapping("student")
3.  public class TestUnifyException {
```

```
4.
5.      @Resource
6.      StudentService studentService;
7.
8.      @GetMapping
9.      public Student getStudent(){
10.         Student student = studentService.getStudent();
11.         return student;
12.     }
13.     @GetMapping("/{id}")
14.     public Student getStudentById(@PathVariable("id") Integer id){
15.         Student student = studentService.getStudentById(id);
16.         return student;
17.     }
18. }
```

在例 11-8 中，存在 getStudent()和 getStudentById()两个方法，在这两个方法中调用了业务层 StudentService 中对应的方法。

在 service 文件夹下创建 StudentService 接口文件，随后在 impl 文件夹下创建该接口的实现类，实现类的代码如例 11-9 所示。

【例 11-9】 StudentServiceImpl.java

```
1.  @Service
2.  public class StudentServiceImpl implements StudentService {
3.
4.      @Override
5.      public Student getStudent() {
6.          String s = null;
7.          s.isEmpty();
8.          return new Student(1,"lisi");
9.      }
10.
11.     @Override
12.     public Student getStudentById(Integer id) {
13.         int a= 2/id;
14.         return new Student(id,"zhangsan");
15.     }
16. }
```

在例 11-9 中，getStudent()方法抛出了空指针异常。getStudentById()方法通过 id 来模拟算术异常。当传入的 id 为 0 时，getStudentById()方法会抛出算术异常。

发布项目，启动 Tomcat，打开浏览器访问 getStudent 业务，结果如图 11.7 所示。

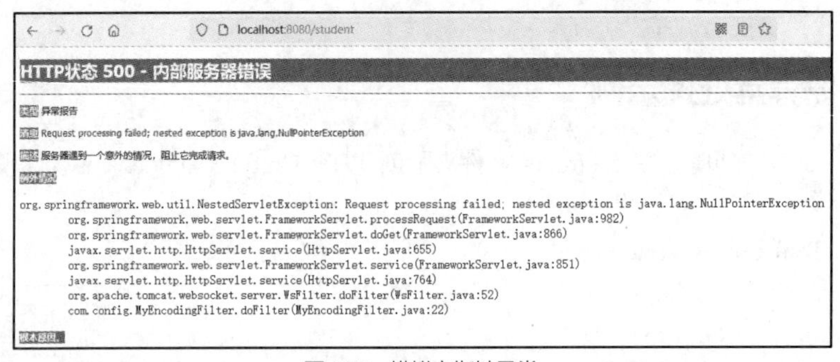

图 11.7　模拟空指针异常

从图 11.7 中可以看出，用户端出现了报错页面。同理，getStudentById 业务也会出现对应异常。此时，系统需要针对不同异常返回不同的信息，把报错页面变为友好的信息页面。下面使用全局异常来解决上述问题。

在 com 文件夹下创建 config 文件夹，在 config 文件夹下新建 MyExceptionHandler.java 文件，其代码如例 11-10 所示。

【例 11-10】 MyExceptionHandler.java

```
1.  @RestControllerAdvice
2.  public class MyExceptionHandler {
3.
4.
5.      @ExceptionHandler(NullPointerException.class)
6.      public Object nullExceptionDealMethod(
7.              HttpServletRequest req,
8.              HttpServletResponse res,
9.              Exception e
10.     ){
11.         System.out.println(res.getContentType());
12.         System.out.println("发生空指针异常");
13.         return new MyResultType(500,"服务器发生错误请联系管理员");
14.     }
15.
16.     @ExceptionHandler(ArithmeticException.class)
17.     public Object arithmeticExceptionDealMethod(
18.             HttpServletRequest req,
19.             HttpServletResponse res,
20.             Exception e
21.     ){
22.         System.out.println("发生算术运算异常");
23.         return new MyResultType(200,"您输入的 id 非法，请重新输入");
24.     }
25.
26. }
```

下面详细介绍例 11-10 中的代码。

（1）在代码的第 1 行中，使用@RestControllerAdvice 注解标明 MyExceptionHandler 类为全局异常管理类，@RestControllerAdvice 注解不但有@ControllerAdvice 注解的功能，其默认还为 MyExceptionHandler 类的所有方法附加@ResponseBody 注解。

（2）在代码的第 5 行中，使用@ExceptionHandler 注解标明 nullExceptionDealMethod()方法为 NullPointerException 异常的处理方法。当程序中发生此异常时，便执行 nullExceptionDealMethod()方法。

（3）在代码的第 16 行中，使用@ExceptionHandler 注解标明 arithmeticExceptionDealMethod()方法为 ArithmeticException 异常的处理方法。当程序中发生此异常时，便执行 arithmeticExceptionDealMethod()方法。

（4）在异常处理方法内可以添加 Exception 属性，在方法内部对异常进行解析。

（5）异常处理方法会把返回值直接返回到前端，在开发中，一般会使用统一的对象作为返回值。在此例中，返回值对象为 MyResultType 类。

下面创建返回值对象，在 com 文件夹下创建 util 文件夹，在 util 文件夹下创建 MyResultType.java 文件，其代码如例 11-11 所示。

【例 11-11】 MyResultType .java

```
1.  public class MyResultType {
```

```
2.     Integer code;
3.     String message;
4.
5.     public MyResultType(String message) {
6.         this.message = message;
7.     }
8.
9.     public MyResultType(Integer code, String message) {
10.        this.code = code;
11.        this.message = message;
12.    }
13.
14.    public MyResultType(Integer code) {
15.        this.code = code;
16.    }
17.
18.    public Integer getCode() {
19.        return code;
20.    }
21.
22.    public void setCode(Integer code) {
23.        this.code = code;
24.    }
25.
26.    public String getMessage() {
27.        return message;
28.    }
29.
30.    public void setMessage(String message) {
31.        this.message = message;
32.    }
33.
34.    @Override
35.    public String toString() {
36.        return "MyResultType{" +
37.                "code=" + code +
38.                ", message='" + message + '\'' +
39.                '}';
40.    }
41. }
```

创建完返回值对象后，运行 Tomcat，因为部分默认 Web 返回数据的编码与浏览器端解析的编码不对应，所以在访问时可能会出现乱码问题。为解决此问题，在 web.xml 中的<context-param>元素下添加如下代码。

```
1.  <filter>
2.      <filter-name>CharacterFilter</filter-name>
3.      <filter-class>
4.          org.springframework.web.filter.CharacterEncodingFilter
5.      </filter-class>
6.      <init-param>
7.          <param-name>encoding</param-name>
8.          <param-value>UTF-8</param-value>
9.      </init-param>
10.     <init-param>
11.         <param-name>forceEncoding</param-name>
```

```
12.         <param-value>true</param-value>
13.     </init-param>
14. </filter>
15. <filter-mapping>
16.     <filter-name>CharacterFilter</filter-name>
17.     <url-pattern>/*</url-pattern>
18. </filter-mapping>
```

重新发布此项目，访问 getStudent 业务，返回结果如图 11.8 所示。

图 11.8 异常处理

从图 11.8 中可以看出，异常得到处理，并返回给客户端一条友好的信息，全局异常配置完毕。

11.5 本章小结

本章首先介绍了 Spring MVC 的文件上传与下载，然后通过一个示例讲解了拦截器的执行流程，接着讲解了 RESTful 风格的实现方法，最后讲解了全局异常处理。其中，RESTful 风格的实现方法是本章的重点内容。

在本章的内容中，RESTful 风格是开发中常用的前后端传输标准，同时也是面试中常涉及的知识点，读者必须熟练掌握。文件上传与下载、拦截器和全局异常处理是开发中偶尔会用到的知识点，读者初步掌握即可。

11.6 习题

1. 填空题

（1）在上传文件时，应该将 ContentType 设置为________。

（2）Spring MVC 中创建的拦截器需要实现________接口。

（3）在 RESTful 风格中，对应增加、删除、修改、查询操作的 Controller 注解分别是________、________、________和________。

2. 选择题

（1）下列关于 Spring MVC 中文件上传和下载的说法正确的是（　　）。

A. 在控制器中，可以使用 MultipartFile 类来接收前端上传的文件

B. 上传文件的数据接口必须使用 POST 请求来传输文件

C. 在上传文件时，需要将上传的表单的 enctype 设置为 multipart/form-data

D. 以上说法全部正确

（2）下列关于拦截器的叙述错误的是（　　）。

A. 拦截器的执行先于过滤器

B. 在自定义拦截器时，preHandle()方法如果返回 false，则此请求不再向后执行

C. 拦截器可以脱离 Servlet 环境独立运行

D. 拦截器是 Spring MVC 提供的，是基于反射生成的，可通过函数调用实现其功能

（3）使用 RESTful 风格进行开发，适配更新操作的注解是（　　）。

A. @PostMapping　　B. @GetMapping

C. @PutMapping　　D. @DeleteMapping

3．思考题

（1）简述 RESTful 风格的实现方法。

（2）简述拦截器与过滤器的区别。

4．编程题

（1）配置一个拦截器，用于获取请求头中的 Host 值。

（2）编写一个全局异常处理类，用于拦截项目中的空指针异常，并向前端返回“发生空指针异常”字符串。

（3）创建 Dog 类，在其中添加 name 和 age 属性。编写一个 RESTful 风格的控制器，用来接收如下的增删改查请求。

```
http://localhost:8080/testDog //以 POST 方式新增，name 和 age 属性在请求体
http://localhost:8080/testDog/1 //以 DELETE 方式删除
http://localhost:8080/testDog //以 PUT 方式更改，id、name 和 age 属性在请求体
http://localhost:8080/testDog/1 //以 GET 方式查找
```

第12章 SSM 框架整合

本章学习目标

- 了解开发中 SSM 框架的目录结构。
- 掌握 MyBatis 配置文件的编写方法。
- 掌握 Maven 工具的使用方法。
- 掌握日志框架的使用方法。

第 2～11 章讲解了 Spring、MyBatis 和 Spring MVC 的使用方法，本章将讲解 Spring、MyBatis 和 Spring MVC 的整合操作，主要是配置文件之间的整合，读者需要尽快熟悉整合后的 SSM 框架开发方式。

12.1 SSM 框架整合概述

SSM 框架整合指的是将 Spring MVC 与 MyBatis 以 Spring 容器的方式进行管理，因为 Spring MVC 是从 Spring 中衍生出来的框架，与 Spring 的兼容性良好，所以 SSM 框架整合主要是将 MyBatis 与 Spring 整合。

SSM 框架整合的方式有 3 种，分别为纯注解方式、纯 XML 方式、注解加 XML 方式。下面的示例中使用注解加 XML 的方式整合 SSM 框架，这种方式是日常开发中常用的方式。

SSM 框架整合的核心是将 MyBatis 交由 Spring 容器进行管理，完成此操作需要进行以下步骤。

（1）将 MyBatis 配置文件中的参数（如数据库配置、数据源配置的相关参数）转移到 Spring 配置文件中。

（2）在 Spring 配置文件中配置 SqlSessionFactory 对象。

下面以一个示例来讲述 SSM 框架整合的方法。

12.2 SSM 框架整合实战

1. 创建数据表

将 Student 对象作为此示例的测试对象，创建数据表的语句如下所示。

```
1.  CREATE TABLE 'student' (
2.    'id' int(0) NOT NULL AUTO_INCREMENT,
3.    'name' varchar(255) ,
4.    PRIMARY KEY ('id') USING BTREE
5.  )
```

```
6.  INSERT INTO 'student' VALUES (1, '钢丹');
7.  INSERT INTO 'student' VALUES (2, '于意');
8.  INSERT INTO 'student' VALUES (3, '李文');
```

2．准备 JAR 包

准备 SSM 框架整合所需的 JAR 包，在此，将 SSM 框架整合常用的 JAR 包全部导入，其中包含 Spring 事务和 SpringAOP 相关的 JAR 包，在使用这些 JAR 包时无须重复导入。需要的 JAR 包如图 12.1 所示。

3．创建 Web 目录结构

SSM 框架整合后的 Web 目录结构相较于整合之前的并没有什么变化，按照 MVC 分层开发模式创建 Web 目录结构，创建完毕的 Web 目录结构如图 12.2 所示。

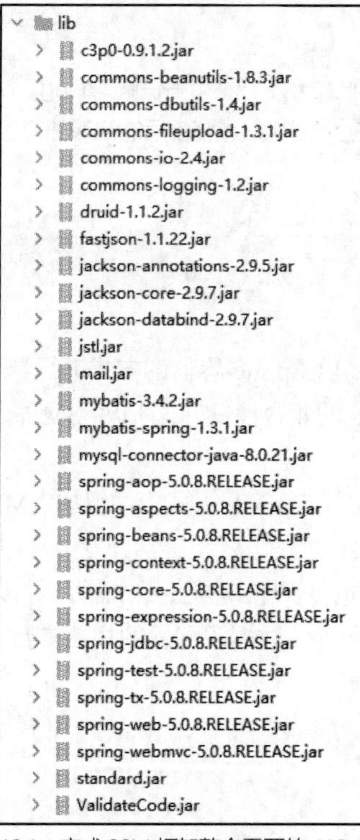

图 12.1　完成 SSM 框架整合需要的 JAR 包

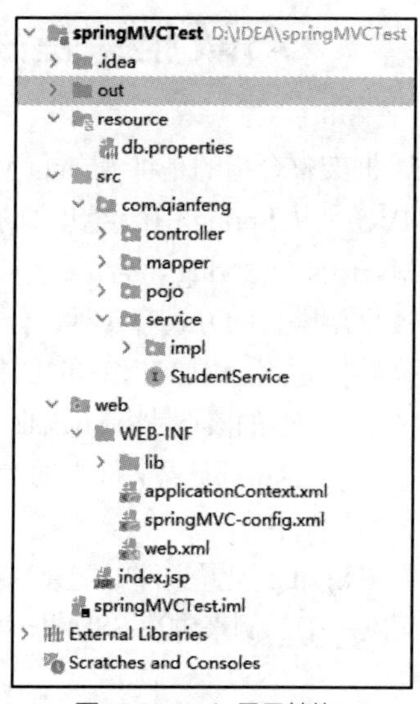

图 12.2　Web 目录结构

4．编写 web.xml 配置文件

在 WEB-INF 文件夹下创建 web.xml 文件，其中的代码如例 12-1 所示。

【例 12-1】 web.xml

```
1.  <!DOCTYPE web-app PUBLIC
2.          "-//Sun Microsystems, Inc.//DTD Web Application 2.3//EN"
3.          "http://java.sun.com/dtd/web-app_2_3.dtd" >
4.  <web-app>
5.
6.      <!--指定 Spring 配置文件的位置-->
7.      <context-param>
8.          <param-name>contextConfigLocation</param-name>
9.          <param-value>WEB-INF/applicationContext.xml</param-value>
```

```
10.     </context-param>
11.
12.     <!--添加过滤器-->
13.     <filter>
14.         <filter-name>CharacterFilter</filter-name>
15.         <filter-class>
16.             org.springframework.web.filter.CharacterEncodingFilter
17.         </filter-class>
18.         <init-param>
19.             <param-name>encoding</param-name>
20.             <param-value>UTF-8</param-value>
21.         </init-param>
22.         <init-param>
23.             <param-name>forceEncoding</param-name>
24.             <param-value>true</param-value>
25.         </init-param>
26.     </filter>
27.     <filter-mapping>
28.         <filter-name>CharacterFilter</filter-name>
29.         <url-pattern>/*</url-pattern>
30.     </filter-mapping>
31.
32.     <!--初始化 IoC 容器，添加监听器-->
33.     <listener>
34.         <listener-class>
35.             org.springframework.web.context.ContextLoaderListener
36.         </listener-class>
37.     </listener>
38.
39.     <!--配置所有请求经过 Spring MVC 的 DispatcherServlet-->
40.     <servlet>
41.         <servlet-name>dispatcher</servlet-name>
42.         <servlet-class>
43.             org.springframework.web.servlet.DispatcherServlet
44.         </servlet-class>
45.         <init-param>
46.             <param-name>contextConfigLocation</param-name>
47.             <param-value>/WEB-INF/springMVC-config.xml</param-value>
48.         </init-param>
49.         <load-on-startup>1</load-on-startup>
50.     </servlet>
51.     <servlet-mapping>
52.         <servlet-name>dispatcher</servlet-name>
53.         <!--拦截所有的 Mapping，即后台 Controller 程序的所有路径-->
54.         <url-pattern>/</url-pattern>
55.     </servlet-mapping>
56.
57.     <!--默认访问首页-->
58.     <welcome-file-list>
59.         <welcome-file>index.jsp</welcome-file>
60.     </welcome-file-list>
61. </web-app>
```

下面详细介绍例 12-1 中的代码。

（1）在初次使用 Spring 时，使用 main()方法通过 ClassPathXmlApplicationContext 类来加载对应的配置文件，由此启动容器 ApplicationContext。在移植到 Web 项目中后，需要通过 Web 容器的启动来加载对应配置文件，因此通过代码的第 33～37 行添加 ContextLoaderListener 监听器来加载 Spring 容器，通过代码的第 7～10 行来指定 Spring 配置文件的位置。

（2）代码的第 40～55 行用于拦截所有请求，使请求经过 Spring MVC，并且指定 Spring 的配置文件。

（3）代码的第 13～30 行用于指定 Web 所用的编码格式。

5．编写 Spring MVC 配置文件

在 WEB-INF 文件夹下创建 springMVC-config.xml 文件，其中的代码如例 12-2 所示。

【例 12-2】 springMVC-config.xml

```
1.  <?xml version="1.0" encoding="UTF-8"?>
2.  <beans xmlns="http://www.springframework.org/schema/beans"
3.         xmlns:xsi="http://www.w3.org/2001/XMLSchema-instance"
4.         xmlns:context="http://www.springframework.org/schema/context"
5.         xmlns:mvc="http://www.springframework.org/schema/mvc"
6.         xsi:schemaLocation="http://www.springframework.org/schema/beans
7.      http://www.springframework.org/schema/beans/spring-beans.xsd
8.      http://www.springframework.org/schema/context
9.      http://www.springframework.org/schema/context/spring-context.xsd
10.     http://www.springframework.org/schema/mvc
11.     http://www.springframework.org/schema/mvc/spring-mvc-3.0.xsd">
12.
13.     <!--开启 Spring MVC 注解模式-->
14.     <mvc:annotation-driven />
15.     <!—配置静态资源访问-->
16.     <mvc:default-servlet-handler/>
17.
18.     <!--          Controller 对象配置块              -->
19.     <!--将控制器类加入 Spring 容器-->
20.     <context:component-scan base-package="com.qianfeng.controller"/>
21.
22.     <!-- 配置视图解析器 -->
23.     <bean class="org.springframework.web.servlet.view.
InternalResourceViewResolver">
24.         <property name="suffix" value=".jsp" />
25.     </bean>
26. </beans>
```

Spring MVC 的配置文件较简单，其主要功能是将 Controller 层的代码扫描到 Spring 容器中。通过第 14 行代码开启 Spring MVC 注解开发模式，通过第 20 行代码将控制器扫描至 Spring 容器中。

6．编写 Spring 配置文件

在 WEB-INF 文件夹下创建 applicationContext.xml 文件，其中的代码如例 12-3 所示。

【例 12-3】 applicationContext.xml

```
1.  <?xml version="1.0" encoding="UTF-8"?>
2.  <beans xmlns="http://www.springframework.org/schema/beans"
3.         xmlns:xsi="http://www.w3.org/2001/XMLSchema-instance"
4.         xmlns:context="http://www.springframework.org/schema/context"
```

```
5.         xmlns:aop="http://www.springframework.org/schema/aop"
6.         xmlns:tx="http://www.springframework.org/schema/tx"
7.         xsi:schemaLocation="http://www.springframework.org/schema/beans
8.            http://www.springframework.org/schema/beans/spring-beans.xsd
9.            http://www.springframework.org/schema/context
10.          http://www.springframework.org/schema/context/spring-context.xsd
11.           http://www.springframework.org/schema/aop
12.           http://www.springframework.org/schema/aop/spring-aop-4.3.xsd
13.           http://www.springframework.org/schema/tx
14.           http://www.springframework.org/schema/tx/spring-tx.xsd">
15.     <!--             数据库连接配置块              -->
16.     <!--引入数据库配置-->
17.     <context:property-placeholder location="classpath:db.properties"/>
18.     <!--创建 DataSource-->
19.     <bean name="dataSource"
20.         class="com.mchange.v2.c3p0.ComboPooledDataSource">
21.        <property name="driverClass" value="${jdbc.myDriver}"/>
22.        <property name="jdbcUrl" value="${jdbc.myUrl}"/>
23.        <property name="user" value="${jdbc.myUsername}"/>
24.        <property name="password" value="${jdbc.myPassword}"/>
25.     </bean>
26.     <!--配置 SqlSession,
27.     此 Bean 创建完成后，就可以通过 Spring 来对数据库进行操作-->
28.     <bean id="sessionFactory"
29.         class="org.mybatis.spring.SqlSessionFactoryBean">
30.        <property name="dataSource" ref="dataSource"/>
31.     </bean>
32.
33.     <!--           Mapper 对象配置块              -->
34.     <!--将 Mapper 类加入 Spring 容器-->
35.     <bean class="org.mybatis.spring.mapper.MapperScannerConfigurer">
36.        <property name="basePackage" value="com.qianfeng.mapper"/>
37.     </bean>
38.
39.     <!--           Service 对象配置块              -->
40.     <!--将业务类加入 Spring 容器-->
41.     <context:component-scan base-package="com.qianfeng.service"/>
42. </beans>
```

下面详细介绍例 12-3 中的代码。

（1）代码的第 17～31 行是原本在 MyBatis 配置文件中的数据库配置，现在移植到 Spring 配置文件中，由 Spring 容器管理。

（2）在 Spring MVC 配置文件中扫描 Controller 层的 Bean 对象，同样，在 Spring 配置文件中需要配置 Service 层与 Mapper 层的 Bean 对象。通过代码的第 35～37 行指定 Mapper 接口的位置，通过代码的第 41 行指定 Service 层的位置。

7．编写 db.properties 配置文件

在 src 文件夹下新建 resource 文件夹，并将其标注为资源文件夹，在其中新建 db.properties 配置文件，其中的代码如下所示。

```
1.   jdbc.myDriver = com.mysql.cj.jdbc.Driver
```

```
2.  jdbc.myUrl = jdbc:mysql://127.0.0.1/test?\
3.    characterEncoding=utf8&\
4.    useSSL=false&\
5.    serverTimezone=UTC&\
6.    allowPublicKeyRetrieval=true
7.  jdbc.myUsername = root
8.  jdbc.myPassword = root
```

8．编写项目代码

在日常开发中，数据库创建完成后，应该按照实体类 POJO 层、Mapper 层、Service 层和 Controller 层的顺序依次编写代码。本例使用 Student 类作为测试对象，按照上述开发顺序编写代码，代码分别如例 12-4、例 12-5、例 12-6 和例 12-7 所示（此处 Service 层的接口代码略写）。

【例 12-4】 Student.java

```
1.  public class Student {
2.      Integer id;
3.      String name;
4.  
5.      public Student() {
6.      }
7.  
8.      public Student(Integer id, String name) {
9.          this.id = id;
10.         this.name = name;
11.     }
12. 
13.     public Integer getId() {
14.         return id;
15.     }
16. 
17.     public void setId(Integer id) {
18.         this.id = id;
19.     }
20. 
21.     public String getName() {
22.         return name;
23.     }
24. 
25.     public void setName(String name) {
26.         this.name = name;
27.     }
28. 
29.     @Override
30.     public String toString() {
31.         return "Student{" +
32.                 "id=" + id +
33.                 ", name='" + name + '\'' +
34.                 '}';
35.     }
36. }
```

【例 12-5】 StudentMapper.java

```
1.  public interface StudentMapper {
2.      @Select("select * from student where id = #{id}")
```

```
3.     Student getStudent(Integer id);
4.  }
```

【例 12-6】 StudentServiceImpl.java

```
1.  @Service
2.  public class StudentServiceImpl implements StudentService {
3.
4.      @Resource
5.      StudentMapper studentMapper;
6.
7.      @Override
8.      public Student getStudentById(Integer id) {
9.          return studentMapper.getStudent(id);
10.     }
11. }
```

【例 12-7】 StudentController.java

```
1.  @RestController
2.  @RequestMapping("student")
3.  public class StudentController {
4.
5.      @Resource
6.      StudentService studentService;
7.
8.      @GetMapping("/{id}")
9.      public Student getStudentById(@PathVariable("id") Integer id){
10.         Student student = studentService.getStudentById(id);
11.         return student;
12.     }
13. }
```

以上的 4 个示例编写了以 Student 类为基础的查询操作代码。下面测试以上示例代码在 SSM 框架整合环境下的运行效果。

9. 测试

发布项目，打开浏览器，访问如下地址。

```
http://localhost:8080/student/1
```

结果如图 12.3 所示。

从图 12.3 中可以看出，该项目的业务流程运行正常，SSM 框架整合完成。

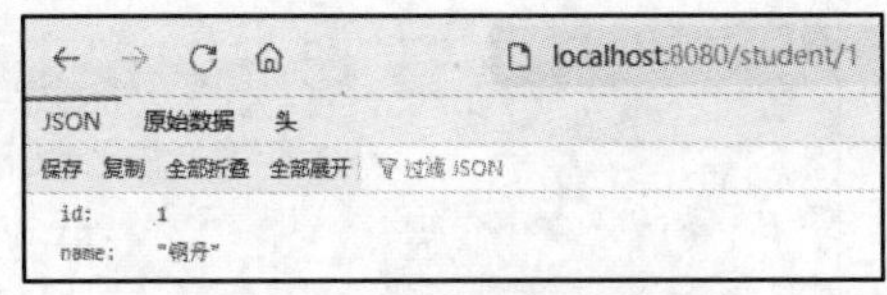

图 12.3　SSM 框架整合测试结果

12.3 整合 Maven 项目

Maven 是一个 JAR 包管理工具，它为 Web 项目提供了一个仓库，这个仓库用来存放所有项目可能用到的 JAR 包。当某个项目需要某个 JAR 包时，只需向 Maven 提供相应 JAR 包的 id，Maven 就会把对应的 JAR 包下载到仓库中并提供给项目。本节将介绍 Maven 项目的整合方法，读者需要熟练掌握本节全部内容。

1．准备 Maven 环境

从官网下载 Maven 包，并将其放在本地磁盘的 maven 文件夹下，在 maven 文件夹下创建 repository 文件夹，此文件夹作为 Maven 的仓库。具体目录结构如图 12.4 所示。

在下载完成的 Maven 包中找到 settings 配置文件，settings 配置文件的位置如图 12.5 所示。

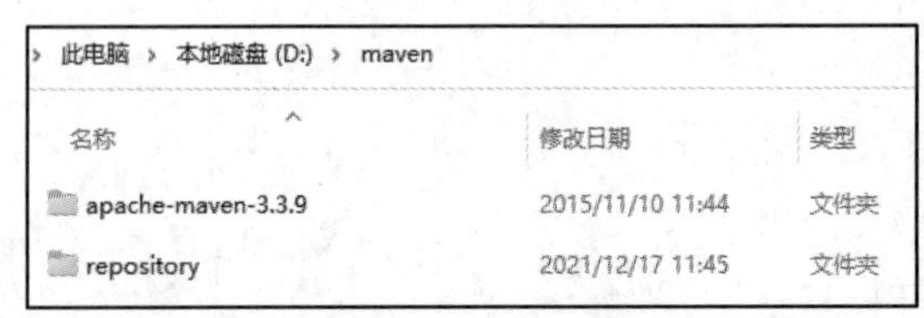

图 12.4　目录结构

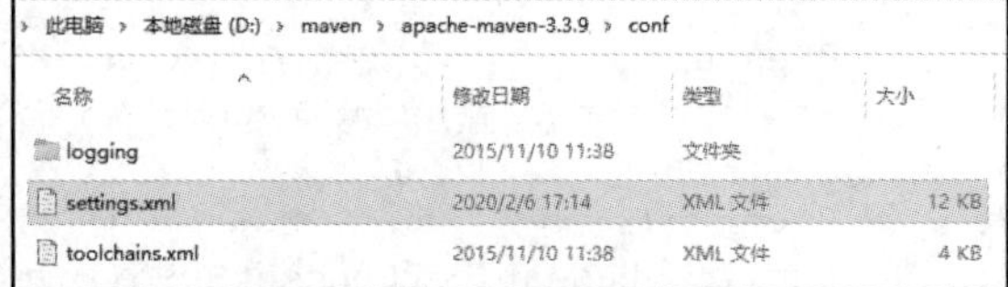

图 12.5　settings 配置文件的位置

打开 settings 配置文件，修改其中的内容。在<settings>元素中添加如下代码来指定仓库的位置。

```
<localRepository>D:/maven/repository</localRepository>
```

在<mirrors>元素中指定镜像仓库，在此可以选择阿里云镜像仓库，具体仓库地址请查阅官方文档。

下面在<profile>元素中指定 Maven 编译的 JDK 版本，具体代码如下所示。

```
1.  <profile>
2.      <id>jdk-1.8</id>
3.      <activation>
4.          <activeByDefault>true</activeByDefault>
5.          <jdk>1.8</jdk>
6.      </activation>
7.      <properties>
8.          <maven.compiler.source>1.8</maven.compiler.source>
9.          <maven.compiler.target>1.8</maven.compiler.target>
10.     <maven.compiler.compilerVersion>1.8</maven.compiler.compilerVersion>
11.     </properties>
12. </profile>
```

2．新建 Maven 项目并配置 Maven 环境

打开 IDEA，新建一个 Maven 项目，如图 12.6 所示。

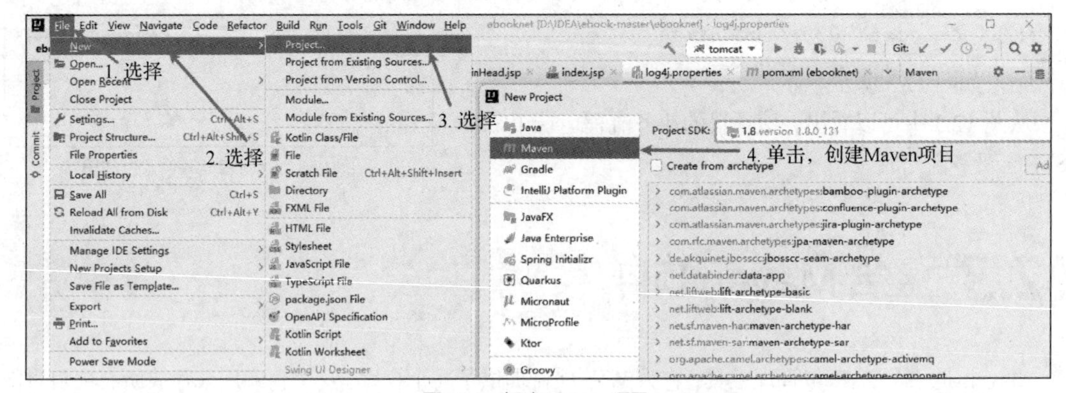

图 12.6　新建 Maven 项目

新建 Maven 项目后，打开“Settings”对话框，在 Maven 设置中配置本地的 Maven 环境，具体操作如图 12.7 所示。

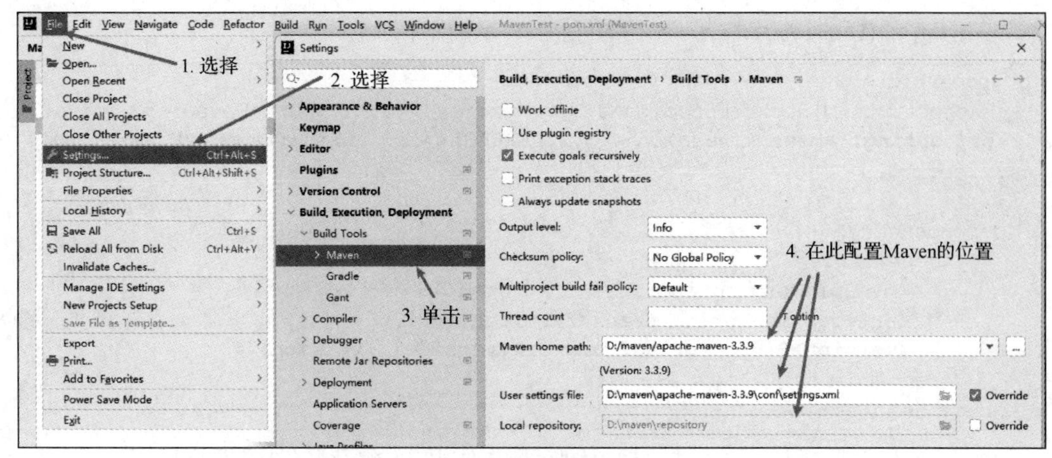

图 12.7 配置 Maven 环境

3．熟悉 Maven 项目的目录结构

新建的 Maven 项目的目录结构如图 12.8 所示。

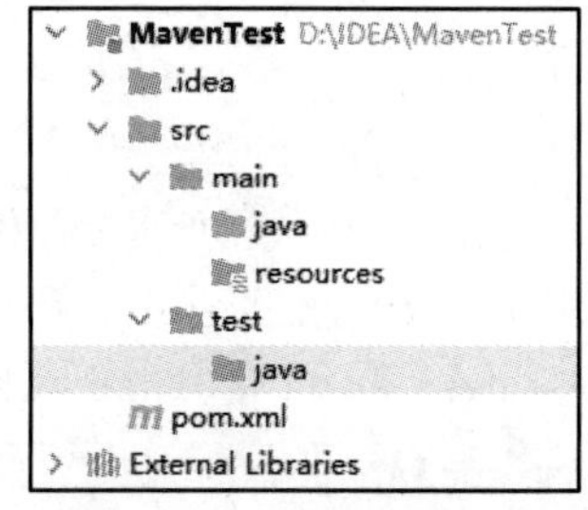

图 12.8 Maven 项目的目录结构

在 Maven 项目的目录结构中，main 文件夹用来存放项目运行所需的文件，test 文件夹用来存放测试文件。在 main 文件夹中，java 文件夹用来存放项目中的代码文件，resources 文件夹用来存放资源文件。pom.xml 文件是 Maven 项目配置所需 JAR 包列表的文件，下面介绍 pom.xml 文件的组成。

4．pom.xml 文件的组成

一个完整的 pom.xml 文件主要包括 4 个部分。下面简单介绍 pom.xml 文件的组成。

（1）第一个部分是配置文件支持的模型版本，代码如下所示。

```
<modelVersion>4.0.0</modelVersion>
```

（2）第二个部分是基本配置，代码如下所示。

```
<groupId>...</groupId>
<artifactId>...</artifactId>
<version>...</version>
<packaging>...</packaging>
<dependencies>...</dependencies>
<dependency>...</dependency>
<parent>...</parent>
<dependencyManagement>...</dependencyManagement>
<modules>...</modules>
<properties>...</properties>
```

在以上代码中，常用的元素是<groupId>元素、<artifactId>元素和<version>元素，将这 3 个元素包裹在<dependency>元素中，可以指定需要的 JAR 包。

在与<dependency>元素结合使用时，<groupId>元素用于指定 JAR 包所在的组名，<artifactId>元素用于指定 JAR 包在组中的标识，<version>元素用于指定 JAR 包的版本。

Maven 会根据<dependency>元素中的信息将相应的 JAR 包下载到 Maven 仓库中并提供给项目使用。在一个 pom.xml 文件中可以引入多个<dependency>元素，将这些元素放在<dependencies>元素中，构成需要引入的 JAR 包列表。

<properties>元素负责抽取代码段，当一段代码需要重复使用时，可以将其放到<properties>元素中，

在需要使用其中的代码时利用“${}”引用即可。具体示例如下所示。

```
1.  <properties>
2.   <project.build.sourceEncoding>UTF-8</project.build.sourceEncoding>
3.   <org.springframework.version>4.3.2.RELEASE</org.springframework.version>
4.  </properties>
5.
6.  <dependencies>
7.      <dependency>
8.          <groupId>org.springframework</groupId>
9.          <artifactId>spring-beans</artifactId>
10.         <version>${org.springframework.version}</version>
11.     </dependency>
12. </dependencies>
```

读者只需初步掌握 Maven 的使用方法。其他元素的使用方法复杂，在此不做讲解。

（3）第三个部分是 Build 配置，代码如下所示。

```
<build>...</build>
<reporting>...</reporting>
```

此部分的代码负责配置和使用 Maven 的插件，<build>元素主要负责配置插件，<reporting>元素负责使用报表插件产生报表规范。

（4）第四个部分是环境配置，代码如下所示。

```
<issueManagement>...</issueManagement>
<ciManagement>...</ciManagement>
<mailingLists>...</mailingLists>
<distributionManagement>...</distributionManagement>
<scm>...</scm>
<prerequisites>...</prerequisites>
<repositories>...</repositories>
<pluginRepositories>...</pluginRepositories>
<profiles>...</profiles>
```

此部分的代码负责配置 Maven 环境，<repositories>元素负责添加远程依赖，<profiles>元素负责构建子模块。其他元素在日常开发中较少使用，读者可以自行查阅其使用方法，本书不过多讲解。

12.4 整合日志框架

在日常开发中，经常需要将某个操作输出，并存放到相应文件夹中。日志框架提供了记录操作并将其持久化的功能。本节将介绍日志框架的使用方法，读者需要深入理解本节内容。

1. 使用 Maven 引入相应的 JAR 包

编写 pom.xml 文件，代码如例 12-8 所示。

【例 12-8】 pom.xml

```
1.  <?xml version="1.0" encoding="UTF-8"?>
2.  <project xmlns="http://maven.apache.org/POM/4.0.0"
3.          xmlns:xsi="http://www.w3.org/2001/XMLSchema-instance"
4.          xsi:schemaLocation="http://maven.apache.org/POM/4.0.0
5.                              http://maven.apache.org/xsd/maven-4.0.0.xsd">
```

```
6.      <modelVersion>4.0.0</modelVersion>
7.
8.      <groupId>org.example</groupId>
9.      <artifactId>MavenTest</artifactId>
10.     <version>1.0-SNAPSHOT</version>
11.
12.     <properties>
13.         <maven.compiler.source>8</maven.compiler.source>
14.         <maven.compiler.target>8</maven.compiler.target>
15.     </properties>
16.     <dependencies>
17.         <!--日志框架-->
18.         <dependency>
19.             <groupId>log4j</groupId>
20.             <artifactId>log4j</artifactId>
21.             <version>1.2.17</version>
22.         </dependency>
23.         <dependency>
24.             <groupId>org.slf4j</groupId>
25.             <artifactId>slf4j-log4j12</artifactId>
26.             <version>1.7.21</version>
27.         </dependency>
28.         <dependency>
29.             <groupId>org.slf4j</groupId>
30.             <artifactId>slf4j-api</artifactId>
31.             <version>1.7.21</version>
32.         </dependency>
33.         <dependency>
34.             <groupId>org.projectlombok</groupId>
35.             <artifactId>lombok</artifactId>
36.             <version>1.18.8</version>
37.         </dependency>
38.     </dependencies>
39. </project>
```

在<dependencies>元素中添加 4 个 JAR 包，其中 lombok 包是简化工具包，其他 3 个包是日志依赖包，将这些 JAR 包引入后，即可进行日志输出。

2. 编写配置文件

在 web.xml 中指定日志框架配置文件的位置，代码如下所示。

```
1.  <!--log4j 配置文件的加载-->
2.  <context-param>
3.      <param-name>log4jConfigLocation</param-name>
4.      <param-value>classpath:config/log4j.properties</param-value>
5.  </context-param>
```

在 resources 文件夹下创建对应的日志框架配置文件，其代码如下所示。

```
1.  log4j.rootLogger =INFO,systemOut,logDailyFile
2.
3.  #输出到控制台
4.  log4j.appender.systemOut = org.apache.log4j.ConsoleAppender
5.  log4j.appender.systemOut.layout = org.apache.log4j.PatternLayout
6.  log4j.appender.systemOut.layout.ConversionPattern =
```

```
7.                  [%-5p][%-22d{yyyy/MM/dd HH:mm:ssS}][%l]%n%m%n
8.  log4j.appender.systemOut.Threshold = INFO
9.  log4j.appender.systemOut.ImmediateFlush = true
10. log4j.appender.systemOut.Target = System.out
11. log4j.logger.org.mybatis=DEBUG
```

第1行代码用于指定将INFO级别以上的日志输出到控制台，第6行代码用于指定输出的格式。

3．使用日志

在需要进行日志输出的类上方标注@Slf4j注解，在此类的方法中，可以直接使用log属性输出相应的信息，代码如下所示。

```
1.  @Slf4j
2.  @ControllerAdvice
3.  public class ExceptionHandlerAdvice {
4.
5.      @ExceptionHandler(Exception.class)
6.      public String allException(Exception exception) {
7.          log.debug("有异常发生");
8.          log.info("有异常发生");
9.          log.warn("有异常发生");
10.         log.error("有异常发生");
11.
12.         return "index";
13.     }
14. }
```

在以上代码中，使用@Slf4j注解标注类，在allException()方法中使用log属性调用相应的日志方法，日志将被输出到配置文件中指定的位置。

12.5 本章小结

本章首先讲解了SSM框架的整合方法，然后讲解了Maven和日志框架的使用方法。在本章内容中，SSM框架的整合是需要重点掌握的部分。此外，Maven是目前企业级开发中主流的JAR包仓库，读者需要熟练掌握Maven的使用方法。

12.6 习题

1．填空题

（1）整合SSM框架的核心是将________交由Spring容器管理。

（2）Maven提供了一个封装________的仓库。

（3）在需要日志输出的类上方标注________注解，然后在此类中，可以直接使用log属性调用输出方法输出响应信息。

2．选择题

（1）下列关于整合 SSM 框架的说法错误的是（　　）。

A. 在整合 SSM 框架的过程中，需要将数据库连接配置由 MyBatis 配置文件移植到 Spring 配置文件中

B. SSM 框架整合的方式有 3 种，分别为纯注解方式、纯 XML 方式、注解加 XML 方式

C. 经过 SSM 框架整合，MyBatis 的相关配置移植到 Spring 配置文件的 SqlSessionFactory 中

D. 在 SSM 框架整合中，MyBatis 配置文件不允许配置

（2）下列关于 SSM 框架整合 Maven 的叙述错误的是（　　）。

A. Maven 的核心是 Settings 配置文件和 Maven 仓库

B. 在配置完 Maven 之后，需要在 IDEA 中指定 Maven 的位置和仓库的位置

C. 在编写 Maven 项目时，需要通过 pom.xml 文件指定项目需要导入的依赖包

D. <dependencies>元素负责进行 Maven 的基本配置

（3）下列 JAR 包中，不是支撑日志框架必需的依赖包是（　　）。

A. log4j　　B. slf4j-log4j　　C. slf4j-api　　D. lombok

3．简述题

（1）简述 SSM 框架整合的要点。

（2）简述 Maven 配置的步骤。

（3）简述 SLF4J 日志框架的使用。

第13章 敛书网 SSM 框架整合项目

本章学习目标

- 了解整合项目中前端页面的编写方法。
- 了解整合项目的设计流程。
- 掌握整合项目中 SSM 框架的搭建方法。
- 掌握整合项目中后端业务代码的编写方法。

本章将详细讲解 SSM 框架整合项目的开发流程。整个项目的开发流程主要包括项目概述、数据库设计、功能代码编写和功能测试。除此之外，此项目对前面所学的知识进行综合运用，其中包括日志管理、MVC 结构和 SSM 框架等核心知识点。本章将带领读者了解敛书网 SSM 框架整合项目的开发流程。

13.1 敛书网项目概述

敛书网是一个基于 SSM 框架的小型实战项目，在此平台中，用户可以完成对书籍的浏览、下载和上传等操作。同时，此项目可以实现管理员账户的设置，通过管理系统的登录，管理员可以控制每个用户上传的书籍，并对用户的反馈信息进行处理。相对于大型项目，此项目可以更好地使读者理解整合项目的架构。本节将带领读者详细了解此项目的功能结构。

13.1.1 功能结构

敛书网的功能结构如图 13.1 所示。

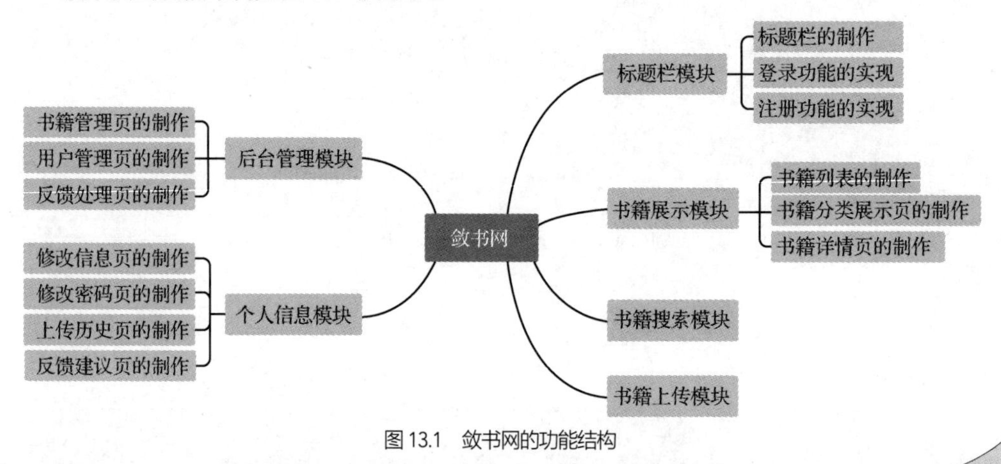

图 13.1　敛书网的功能结构

从图 13.1 中可以看出，敛书网项目分为六大模块，分别为标题栏模块、书籍展示模块、书籍搜索模块、书籍上传模块、个人信息模块和后台管理模块。每个模块都有具体的功能实现，下面讲解每个模块的具体功能。

13.1.2 功能预览

敛书网由多个页面组成，在浏览器中访问敛书网的首页，效果如图 13.2 所示。

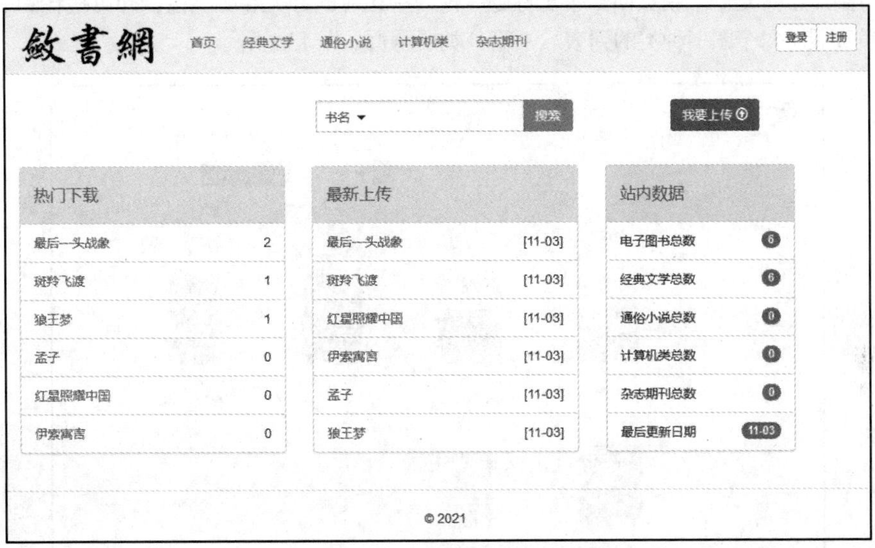

图 13.2 敛书网的首页效果

从图 13.2 中可以看出，敛书网的首页由 4 个模块组成，这些模块分别是标题栏模块、书籍搜索模块、书籍上传模块和书籍展示模块。此外，当用户登录后，标题栏右侧会展示个人信息模块。下面将详细讲解这些模块的页面组成和功能。

1．标题栏模块

标题栏模块分为标题 logo、书籍分类导航链接和登录注册按钮组 3 个组件。标题 logo 处于标题栏的左侧。书籍分类导航链接在标题栏的中间，单击每个分类导航链接即可跳转到相应的分类页面。登录注册按钮组处于标题栏的右侧，单击登录或注册按钮，可以弹出相应的登录或注册对话框，分别如图 13.3 和图 13.4 所示。

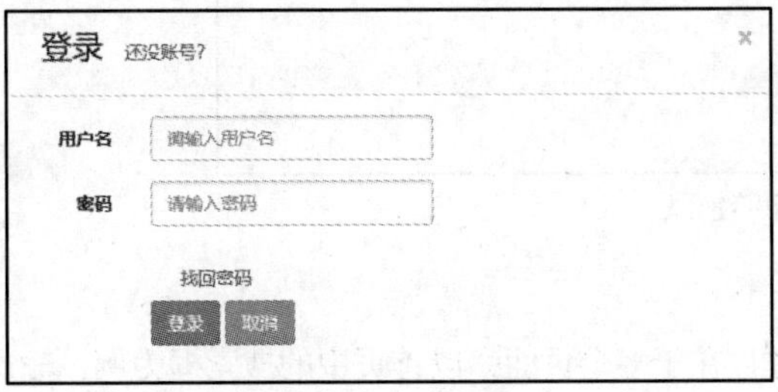

图 13.3 登录对话框

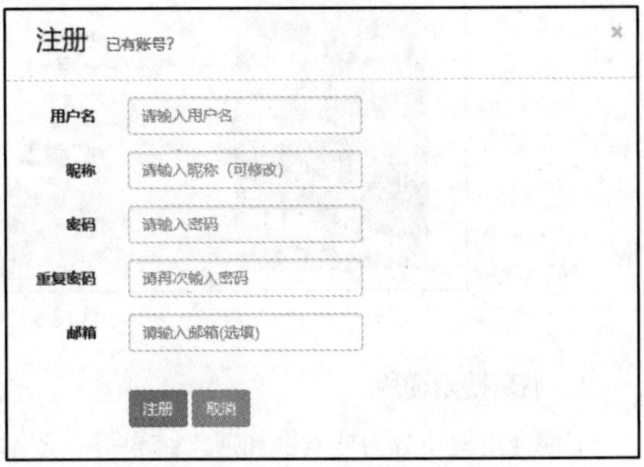

图 13.4 注册对话框

在登录对话框内输入相应的信息后，单击“登录”按钮即可完成登录。注册对话框中的操作类似。

2．书籍展示模块

书籍展示模块分为书籍列表、书籍分类展示页和书籍详情页 3 个部分。

书籍列表处于首页的中央，分为热门下载、最新上传和站内数据 3 个部分。其中，热门下载展示的是图书下载次数的排行，最新上传展示的是最近上传的图书，站内数据展示的是各个分类的图书总数。

书籍分类展示页是一个新页面，单击标题栏模块中相应的分类导航链接，即可跳转到书籍分类展示页，它负责展示每个图书种类的列表。书籍分类展示页如图 13.5 所示。

图 13.5　书籍分类展示页

书籍详情页负责展示图书的具体信息，跳转到此页面有多种方法，在此介绍常用的两种方法。第一种方法是单击书籍分类展示页中图书的“详情”按钮，第二种方法是单击首页热门下载和最新上传部分的图书链接。书籍详情页如图 13.6 所示。

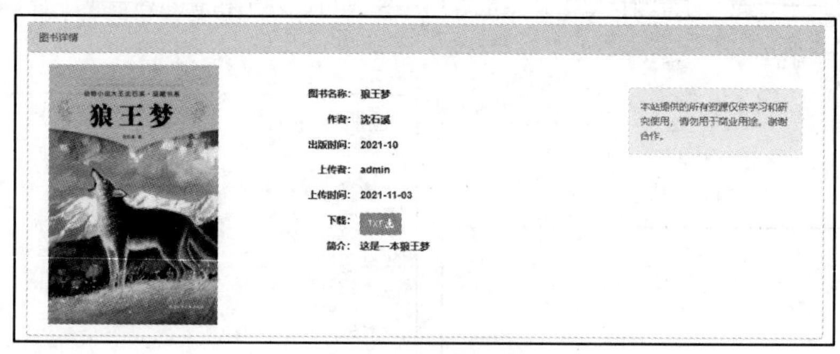

图 13.6　书籍详情页

3．书籍搜索模块

书籍搜索模块分为搜索框和搜索结果页。搜索框存在于多个页面，以首页中的搜索框为例，输入需要搜索的图书名，即可跳转到搜索结果页。搜索结果页如图 13.7 所示。

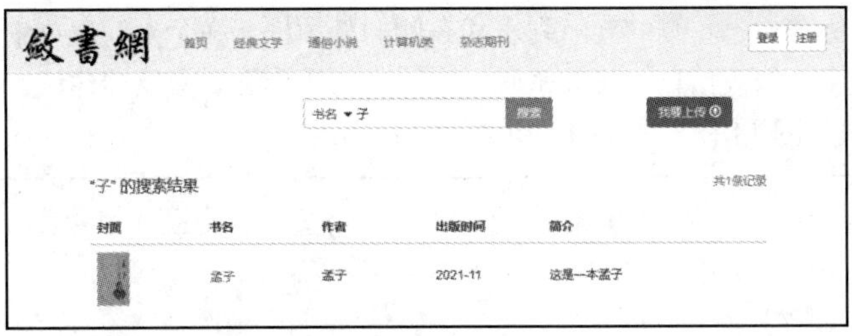

图 13.7 搜索结果页

在搜索结果页单击每本书的书名即可跳转到相应的书籍详情页。

4．书籍上传模块

书籍上传模块存在于多个页面，以首页的“我要上传”按钮为例，单击此按钮后（需要登录），即可跳转到书籍上传页。书籍上传页如图 13.8 所示。

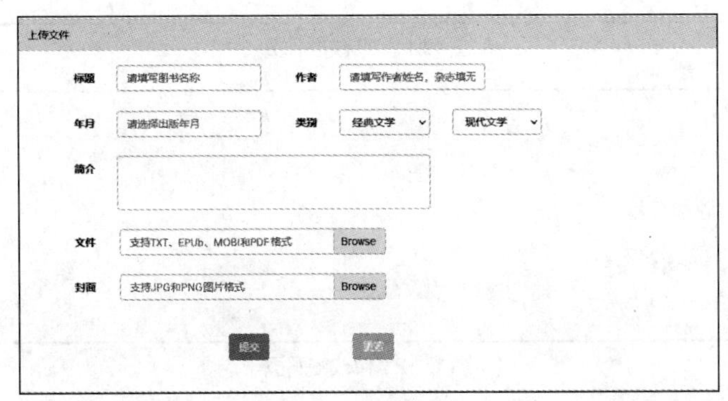

图 13.8 书籍上传页

在图 13.8 所示的页面中，填写上传图书的信息，选择相应的图书文件，单击“提交”按钮即可上传图书。

5．个人信息模块

个人信息模块用于查看和修改个人信息（需要登录）。单击首页标题栏中的头像，在弹出的快捷框中单击“个人资料”即可跳转到个人信息页。个人信息模块如图 13.9 所示，个人信息页如图 13.10 所示。

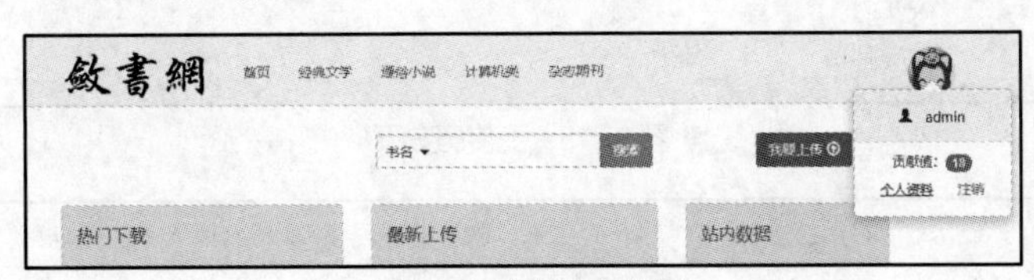

图 13.9 个人信息模块

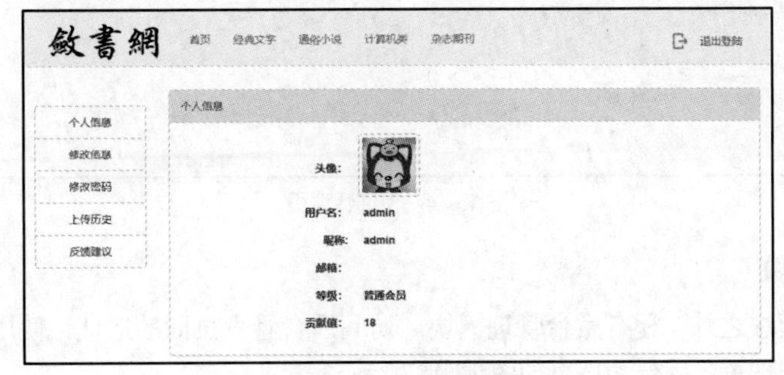

图 13.10 个人信息页

个人信息页分为标题栏、左侧的导航列表和右侧的展示框。在个人信息页中，通过左侧的导航列表，可以查看个人信息、修改信息、修改密码、查看上传历史和提交反馈建议，对应的各个页面的预览效果分别如图 13.11、图 13.12、图 13.13 和图 13.14 所示。

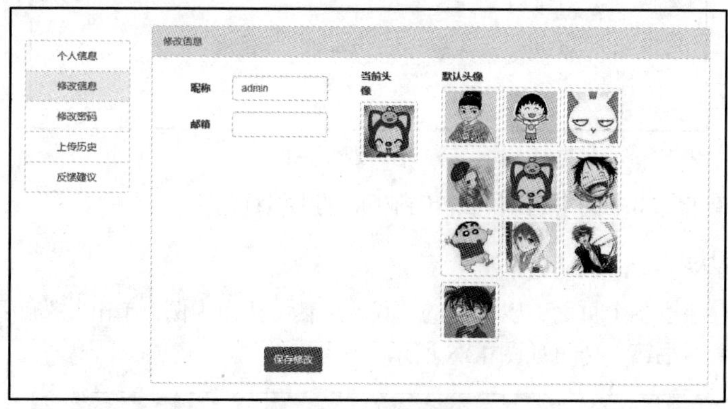

图 13.11　修改信息页

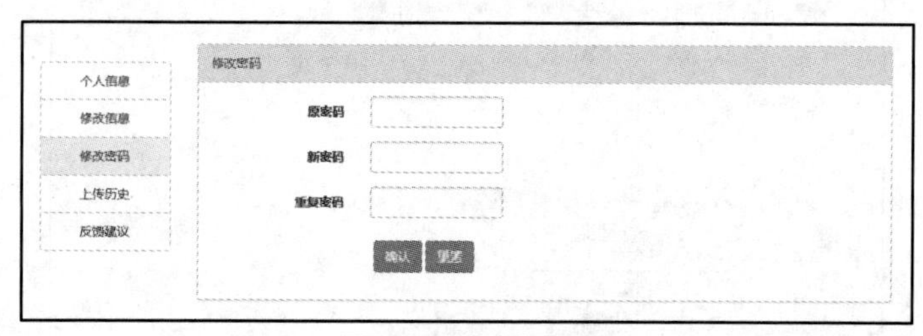

图 13.12　修改密码页

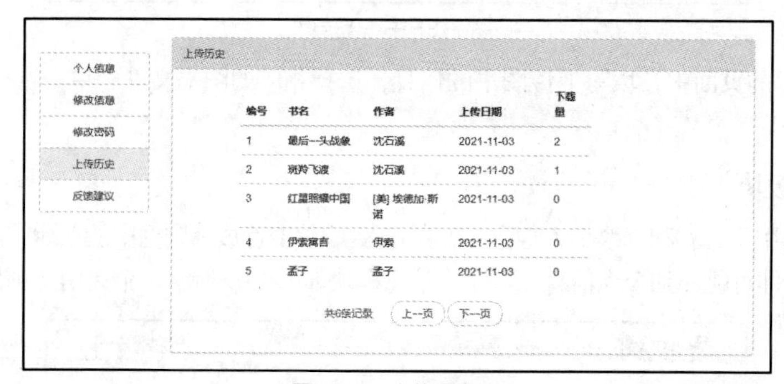

图 13.13　上传历史页

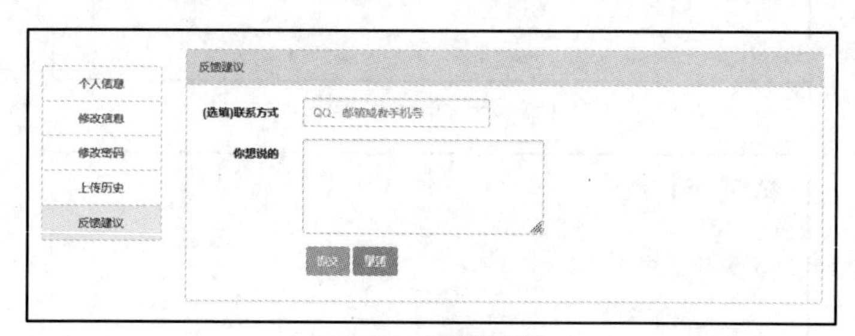

图 13.14　反馈建议页

6．后台管理模块

敛书网除上述模块之外，还有后台管理模块。访问后台管理页面的地址，使用给定的账户登录系统后即可进入后台管理页。后台管理页如图 13.15 所示。

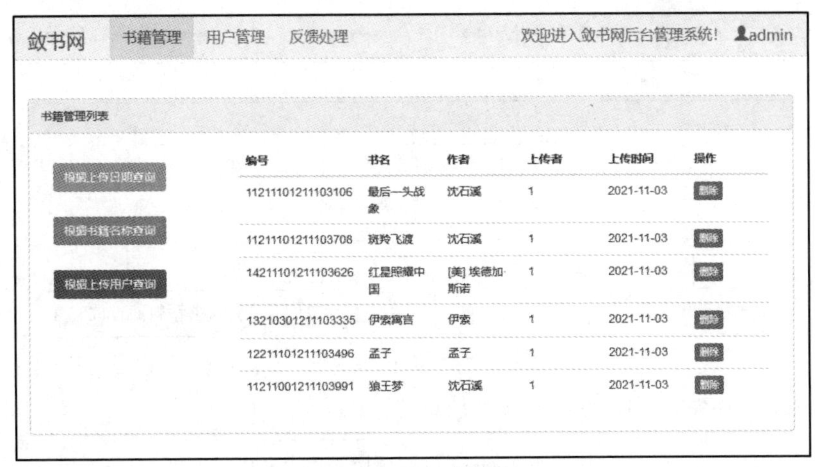

图 13.15　后台管理页

从图 13.15 中可以看出，后台管理模块具有书籍管理、用户管理和反馈处理 3 个功能。其中书籍管理页如图 13.15 所示，用户管理页和反馈处理页分别如图 13.16 和图 13.17 所示。

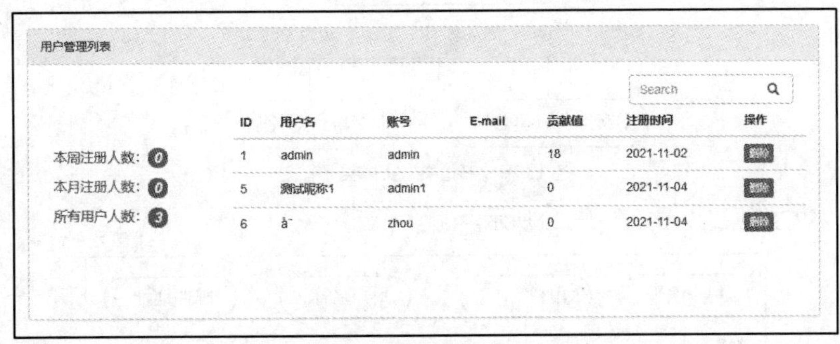

图 13.16　用户管理页

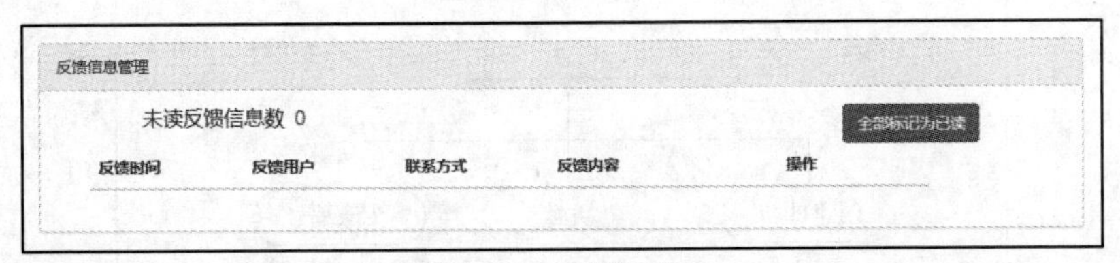

图 13.17　反馈处理页

13.2 数据库设计

在开发应用程序时，需要先确定项目中的实体类，然后根据实体类之间的关系创建数据表。本节将带领读者完成数据库的相关设计。

13.2.1 设计 E-R 图

在本小节介绍一种描述实体类对象关系的模型：E-R 图。E-R 图又称为实体-关系图，它能够直观地描述实体与属性之间的关系。下面根据敛书网的功能来设计 E-R 图，具体如下所示。

（1）书籍实体类的 E-R 图如图 13.18 所示。

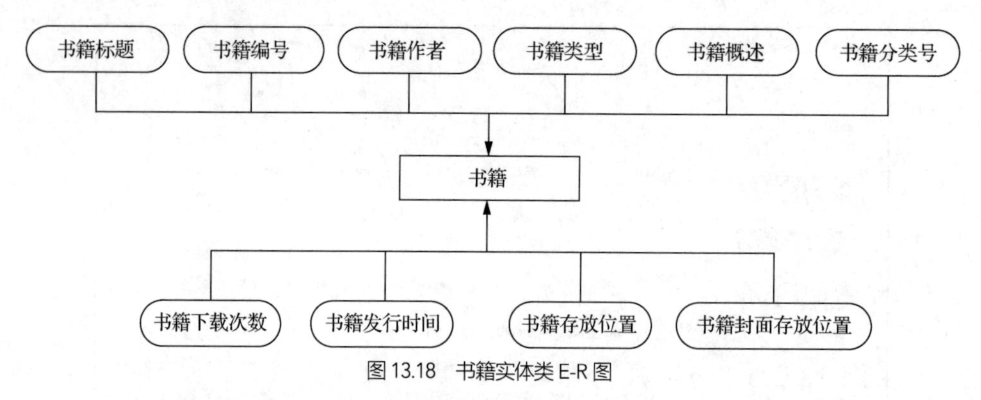

图 13.18　书籍实体类 E-R 图

（2）书籍类型实体类的 E-R 图如图 13.19 所示。

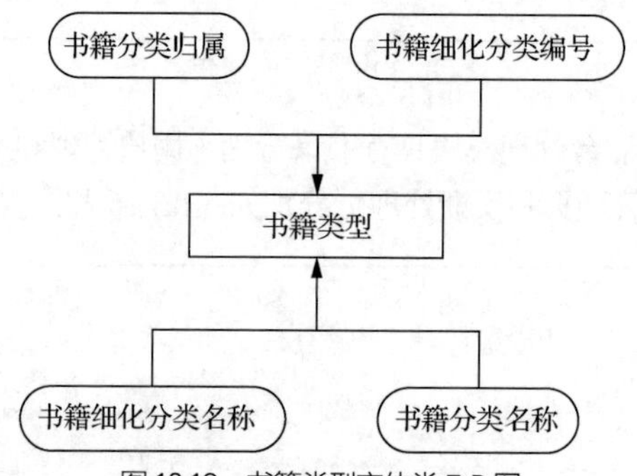

图 13.19　书籍类型实体类 E-R 图

（3）用户实体类的 E-R 图如图 13.20 所示。

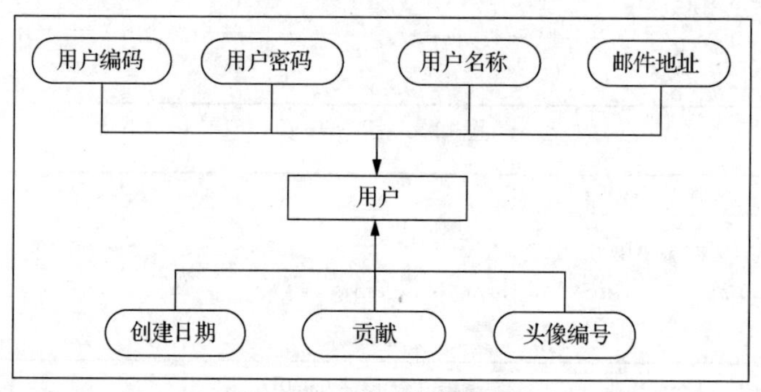

图 13.20　用户实体类 E-R 图

（4）书籍上传实体类的 E-R 图如图 13.21 所示。

（5）反馈建议实体类的 E-R 图如图 13.22 所示。

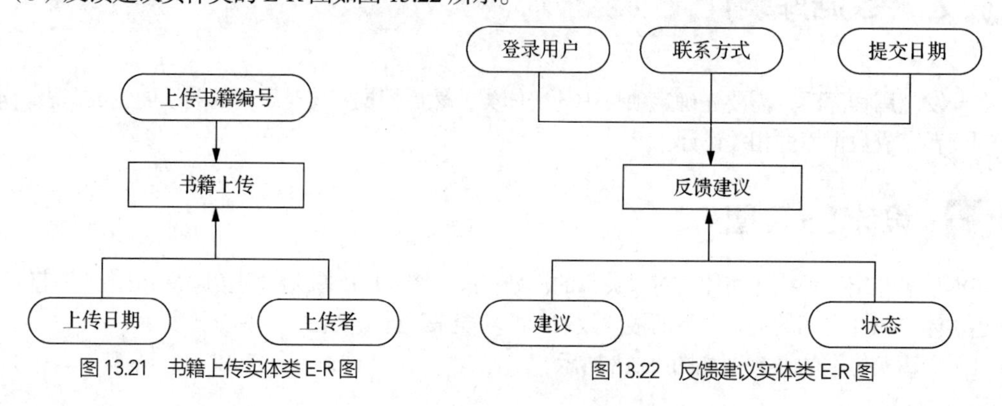

图 13.21　书籍上传实体类 E-R 图

图 13.22　反馈建议实体类 E-R 图

（6）贡献实体类的 E-R 图如图 13.23 所示。

（7）头像实体类的 E-R 图如图 13.24 所示。

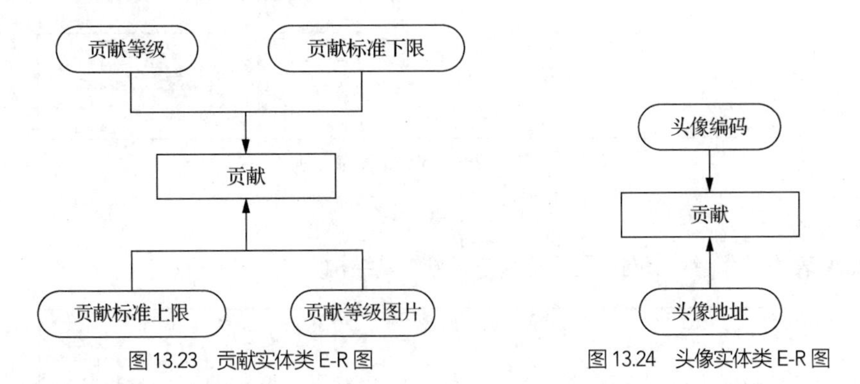

图 13.23　贡献实体类 E-R 图　　图 13.24　头像实体类 E-R 图

13.2.2 设计数据表

在了解每个实体类的 E-R 图之后，根据每个 E-R 图创建数据表。

1．book 表

book 表用于存储敛书网中所有的书籍，其结构如图 13.25 所示。

名	类型	长度	小数点	不是 null	虚拟	键	注释
id	bigint	0	0	☑	☐	1	书籍编号
book_title	varchar	50	0	☑	☐		书籍标题
book_author	varchar	30	0	☑	☐		书籍作者
book_summary	text	0	0	☐	☐		书籍概述
type_id	int	0	0	☐	☐		书籍分类号
download_times	int	0	0	☐	☐		书籍下载次数
book_pubYear	date	0	0	☐	☐		书籍发行时间
book_file	varchar	100	0	☐	☐		书籍存放位置
book_cover	varchar	100	0	☐	☐		书籍封面存放位置
book_format	varchar	10	0	☑	☐		书籍类型

图 13.25　book 表的结构

2．book_type 表

book_type 表用于存储敛书网中书籍的类型，其结构如图 13.26 所示。

名	类型	长度	小数点	不是 null	虚拟	键	注释
id	int	0	0	☑	☐	1	书籍类型id
book_large_type	int	0	0	☑	☐		书籍分类归属
book_small_type	int	0	0	☑	☐		书籍细化分类编号
large_type_name	varchar	20	0	☑	☐		书籍分类名称
small_type_name	varchar	20	0	☑	☐		书籍细化分类名称

图 13.26　book_type 表的结构

3．user 表

user 表用于存储敛书网中所有的用户信息，其结构如图 13.27 所示。

名	类型	长度	小数点	不是 null	虚拟	键	注释
id	bigint	0	0	☑	☐	1	系统自增主键
userCode	varchar	20	0	☑	☐		用户编码
userPassword	varchar	20	0	☑	☐		用户密码
userName	varchar	20	0	☑	☐		用户名称
email	varchar	30	0	☐	☐		邮件地址
avatarNum	smallint	0	0	☐	☐		头像编号
contribution	int	0	0	☐	☐		贡献
creationDate	date	0	0	☐	☐		创建日期

图 13.27　user 表的结构

4．upload 表

upload 表用于存储敛书网中用户上传的图书的信息，其结构如图 13.28 所示。

名	类型	长度	小数点	不是 null	虚拟	键	注释
id	int	0	0	☑	☐	1	系统自增主键
uploader	bigint	0	0	☑	☐		上传者
uploadedBook	bigint	0	0	☑	☐		上传书籍编号
uploadedDate	date	0	0	☐	☐		上传日期

图 13.28　upload 表的结构

5．feedback 表

feedback 表用于存储敛书网中用户的反馈信息，其结构如图 13.29 所示。

名	类型	长度	小数点	不是 null	虚拟	键	注释
id	int	0	0	☑	☐	1	系统自增主键
loginedUser	bigint	0	0	☐	☐		登录用户
contact	varchar	30	0	☐	☐		联系方式
suggestion	varchar	500	0	☐	☐		建议
status	int	0	0	☑	☐		状态
postTime	timestamp	0	0	☑	☐		提交日期

图 13.29　feedback 表的结构

6．contribution 表

contribution 表用于存储敛书网中用户贡献书籍等级的信息，其结构如图 13.30 所示。

名	类型	长度	小数点	不是 null	虚拟	键	注释
id	smallint	0	0	☑	☐	1	系统自增主键
level_txt	varchar	10	0	☐	☐		贡献等级
lowerLimit	int	0	0	☐	☐		贡献标准下限
upperLimit	int	0	0	☐	☐		贡献标准上限
level_img	varchar	100	0	☐	☐		贡献等级图片

图 13.30　contribution 表的结构

7．avatar 表

avatar 表用于存储敛书网中用户头像的信息，其结构如图 13.31 所示。

名	类型	长度	小数点	不是 null	虚拟	键	注释
id	smallint	0	0	☑	☐	1	头像id
avatar_txt	varchar	20	0	☐	☐		头像编码
avatar_img	varchar	100	0	☐	☐		头像地址

图 13.31　avatar 表的结构

创建完上述数据表后，向表中添加样例数据，样例数据可参考项目源码包中的 ebook.sql 文件。

13.3 项目搭建

在开发功能之前，首先要进行项目搭建工作。本节将分步讲解整合项目的搭建过程。

项目的开发环境及所需工具如下。

- Web 服务器：Tomcat8。
- Java 开发包：JDK1.8。
- 数据库：MySQL8.0。
- 开发工具：IDEA2021.2。
- 浏览器：Firefox。

13.3.1　创建 Maven 项目

打开 IDEA，创建 Maven 项目，并将其命名为 ebooknet，创建完成的项目如图 13.32 所示。

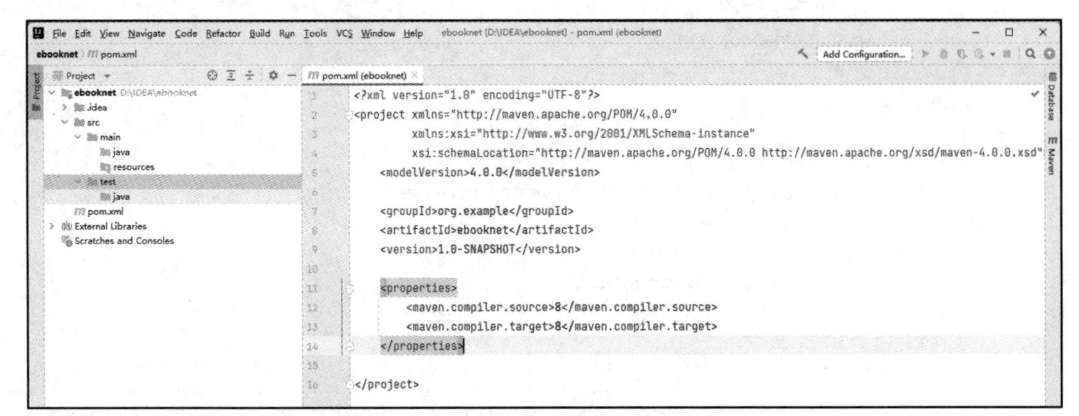

图 13.32　创建的 Maven 项目

在 pom.xml 文件中配置需要的依赖。此项目要用到的依赖如下所示。

（1）Spring 的相关依赖，其代码如下所示。

```
1.    <dependency>
2.       <groupId>org.springframework</groupId>
3.       <artifactId>spring-beans</artifactId>
4.       <version>${org.springframework.version}</version>
5.    </dependency>
6.    <dependency>
7.       <groupId>org.springframework</groupId>
8.       <artifactId>spring-context</artifactId>
9.       <version>${org.springframework.version}</version>
10.   </dependency>
11.   <dependency>
12.      <groupId>org.springframework</groupId>
13.      <artifactId>spring-core</artifactId>
14.      <version>${org.springframework.version}</version>
15.   </dependency>
16.   <dependency>
17.      <groupId>org.springframework</groupId>
18.      <artifactId>spring-expression</artifactId>
19.      <version>${org.springframework.version}</version>
20.   </dependency>
21.   <dependency>
22.      <groupId>org.springframework</groupId>
23.      <artifactId>spring-web</artifactId>
24.      <version>${org.springframework.version}</version>
25.   </dependency>
26.   <dependency>
27.      <groupId>org.springframework</groupId>
28.      <artifactId>spring-webmvc</artifactId>
29.      <version>${org.springframework.version}</version>
30.   </dependency>
```

以上代码中包含了 Spring 和 Spring MVC 的基础依赖包。其中，<version>元素使用“${}”引用<properties>元素中设定的版本号，此处 Spring 的版本号为 4.3.2.RELEASE。

（2）数据库的依赖，其代码如下所示。

```
1.    <dependency>
2.        <groupId>mysql</groupId>
3.        <artifactId>mysql-connector-java</artifactId>
4.        <version>8.0.13</version>
5.    </dependency>
6.    <dependency>
7.        <groupId>org.springframework</groupId>
8.        <artifactId>spring-jdbc</artifactId>
9.        <version>${org.springframework.version}</version>
10.   </dependency>
11.   <dependency>
12.       <groupId>com.alibaba</groupId>
13.       <artifactId>druid</artifactId>
14.       <version>1.0.29</version>
15.   </dependency>
```

以上代码的第 1～5 行采用了 8.0.13 版本的数据库连接依赖，这个依赖是 8.0 以后版本的 MySQL 与 Spring 连接所需的依赖。如果是 8.0 版本以前的 MySQL，则此处需要使用 8.0 以前的 MySQL 连接依赖。

（3）日志框架的依赖，其代码如下所示。

```
1.    <dependency>
2.        <groupId>log4j</groupId>
3.        <artifactId>log4j</artifactId>
4.        <version>1.2.17</version>
5.    </dependency>
6.    <dependency>
7.        <groupId>org.slf4j</groupId>
8.        <artifactId>slf4j-log4j12</artifactId>
9.        <version>1.7.21</version>
10.   </dependency>
11.   <dependency>
12.       <groupId>org.slf4j</groupId>
13.       <artifactId>slf4j-api</artifactId>
14.       <version>1.7.21</version>
15.   </dependency>
16.   <dependency>
17.       <groupId>org.projectlombok</groupId>
18.       <artifactId>lombok</artifactId>
19.       <version>1.18.8</version>
20.   </dependency>
```

添加以上代码，引入日志框架所需的依赖。此处选用 log4j 作为日志支持，选用 slf4j 作为日志门面，并配合 lombok 依赖实现@slf4j 注解便捷的日志标注。

（4）Spring 事务的依赖，其代码如下所示。

```
1.    <dependency>
2.        <groupId>org.springframework</groupId>
3.        <artifactId>spring-tx</artifactId>
4.        <version>${org.springframework.version}</version>
5.    </dependency>
6.    <dependency>
7.        <groupId>org.aspectj</groupId>
8.        <artifactId>aspectjweaver</artifactId>
9.        <version>1.8.9</version>
```

```
10.  </dependency>
11.  <dependency>
12.      <groupId>aspectj</groupId>
13.      <artifactId>aspectjrt</artifactId>
14.      <version>1.5.3</version>
15.  </dependency>
```

添加以上代码，引入 Spring 事务所需的依赖，并引入 aspectJ 作为 Spring AOP 的支持。

（5）Spring 测试的依赖，其代码如下所示。

```
1.   <dependency>
2.       <groupId>org.springframework</groupId>
3.       <artifactId>spring-test</artifactId>
4.       <version>${org.springframework.version}</version>
5.   </dependency>
6.   <dependency>
7.       <groupId>org.testng</groupId>
8.       <artifactId>testng</artifactId>
9.       <version>6.9.10</version>
10.  </dependency>
11.  <dependency>
12.      <groupId>org.hamcrest</groupId>
13.      <artifactId>hamcrest-core</artifactId>
14.      <version>1.3</version>
15.  </dependency>
16.  <dependency>
17.      <groupId>org.hamcrest</groupId>
18.      <artifactId>hamcrest-library</artifactId>
19.      <version>1.3</version>
20.  </dependency>
```

在 Spring 项目中需要使用 Spring 测试功能，因此，添加以上代码，引入 Spring 测试所需的依赖。

（6）文件上传的依赖，其代码如下所示。

```
1.   <dependency>
2.       <groupId>commons-fileupload</groupId>
3.       <artifactId>commons-fileupload</artifactId>
4.       <version>1.3.1</version>
5.   </dependency>
6.   <dependency>
7.       <groupId>commons-io</groupId>
8.       <artifactId>commons-io</artifactId>
9.       <version>2.4</version>
10.  </dependency>
```

因为敛书网涉及大量的书籍上传操作，所以添加以上代码，引入文件上传所需的依赖。

（7）Servlet 依赖，其代码如下所示。

```
1.   <dependency>
2.       <groupId>javax.servlet</groupId>
3.       <artifactId>servlet-api</artifactId>
4.       <version>2.5</version>
5.   </dependency>
6.   <dependency>
7.       <groupId>javax.servlet.jsp</groupId>
8.       <artifactId>javax.servlet.jsp-api</artifactId>
9.       <version>2.3.1</version>
```

```
10.   </dependency>
11.   <dependency>
12.       <groupId>javax.servlet</groupId>
13.       <artifactId>jstl</artifactId>
14.       <version>1.2</version>
15.   </dependency>
16.   <dependency>
17.       <groupId>taglibs</groupId>
18.       <artifactId>standard</artifactId>
19.       <version>1.1.2</version>
20.   </dependency>
```

Servlet 依赖是项目必须引入的依赖，因此添加以上代码，引入 Web 基础支持。

（8）MyBatis 依赖，其代码如下所示。

```
1.    <dependency>
2.        <groupId>org.mybatis</groupId>
3.        <artifactId>mybatis</artifactId>
4.        <version>3.4.2</version>
5.    </dependency>
6.    <dependency>
7.        <groupId>org.mybatis</groupId>
8.        <artifactId>mybatis-spring</artifactId>
9.        <version>1.3.0</version>
10.   </dependency>
```

此项目整合了 MyBatis，因此添加以上代码，引入 MyBatis 依赖。

（9）序列化的依赖，其代码如下所示。

```
1.    <dependency>
2.        <groupId>com.fasterxml.jackson.core</groupId>
3.        <artifactId>jackson-core</artifactId>
4.        <version>2.8.6</version>
5.    </dependency>
6.    <dependency>
7.        <groupId>com.fasterxml.jackson.core</groupId>
8.        <artifactId>jackson-annotations</artifactId>
9.        <version>2.8.6</version>
10.   </dependency>
11.   <dependency>
12.       <groupId>com.fasterxml.jackson.core</groupId>
13.       <artifactId>jackson-databind</artifactId>
14.       <version>2.8.6</version>
15.   </dependency>
```

在敛书网项目中，实体类使用序列化，因此添加以上代码，引入序列化相关的依赖。除此之外，还需要配置 Maven 的 Resource 文件过滤等，具体代码见项目源码包中的 pom.xml 文件。

13.3.2 搭建 SSM 框架

在 Maven 项目中，创建 SSM 框架的目录结构。创建完成的目录结构如图 13.33 所示。

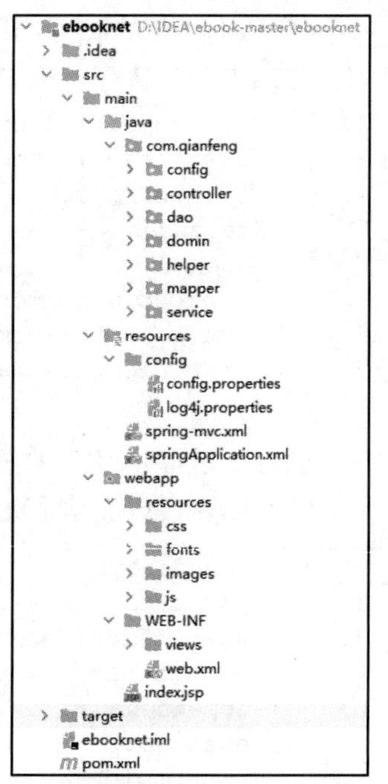

图 13.33　目录结构

在图 13.33 中，java 文件夹下存放项目的核心代码，与其同级的

resources 文件夹下存放配置文件，webapp 文件夹下存放静态资源文件，WEB-INF 文件夹下存放 JSP 文件和 web.xml 文件。

在此将项目所需的实体类和前端资源文件引入。实体类文件在项目源码包的 helper 和 domin 文件夹下。前端资源文件在项目源码包中的 resources 文件夹下，需要注意的是，此 resources 文件夹处于 webapp 文件夹下。下面编写项目所需的配置文件。

在与 java 文件夹同级的 resources 文件夹下创建 Spring 配置文件，并将其命名为 springApplication.xml。编写 Spring 配置文件的代码，如例 13-1 所示。

【例 13-1】 springApplication.xml

```
1.  <?xml version="1.0" encoding="utf-8"?>
2.  <beans xmlns="http://www.springframework.org/schema/beans"
3.        xmlns:xsi="http://www.w3.org/2001/XMLSchema-instance"
4.        xmlns:context="http://www.springframework.org/schema/context"
5.        xsi:schemaLocation="http://www.springframework.org/schema/beans
6.          http://www.springframework.org/schema/beans/spring-beans.xsd
7.          http://www.springframework.org/schema/context
8.         http://www.springframework.org/schema/context/spring-context.xsd">
9.
10.     <context:component-scan base-package="com.qianfeng" />
11.
12.     <!--引入数据库配置-->
13.     <context:property-placeholder
14.           location="classpath:config/config.properties"/>
15.
16.     <!-- 配置数据源，使用阿里巴巴的 Druid 数据源 -->
17.     <bean id="dataSource" class="com.alibaba.druid.pool.DruidDataSource">
18.        <property name="driverClassName" value="${jdbc.driver}" />
19.        <property name="url" value="${jdbc.url}" />
20.        <property name="username" value="${jdbc.username}" />
21.        <property name="password" value="${jdbc.password}" />
22.     </bean>
23.
24.     <!-- 配置 SqlSessionFactory -->
25.     <bean id="sqlSessionFactory"
26.          class="org.mybatis.spring.SqlSessionFactoryBean">
27.        <property name="dataSource" ref="dataSource"/>
28.        <property name="typeAliasesPackage" value="com.qianfeng.domin"/>
29.        <!-- 配置 MyBatis 文件路径 -->
30.        <property name="mapperLocations"
31.                value="classpath*:com/qianfeng/mapper/*.xml" />
32.     </bean>
33.
34.     <bean id="mapperScannerConfigurer"
35.          class="org.mybatis.spring.mapper.MapperScannerConfigurer" >
36.        <property name="basePackage" value="com.qianfeng.dao" />
37.     </bean>
38.
39. </beans>
```

在例 13-1 中配置了数据库及 MyBatis 相关的设置。下面创建 Spring 配置文件所需的资源文件。在与 java 文件夹同级的 resources 文件夹下创建 config 文件夹，在 config 文件夹下创建 config.properties 配置文件并在其中添加如下代码。

```
1.  # MySQL 数据库配置信息
2.  jdbc.driver=com.mysql.cj.jdbc.Driver
3.  jdbc.url=jdbc:mysql://127.0.0.1/ebook?\
4.    characterEncoding=utf8&\
5.    useSSL=false&\
6.    serverTimezone=UTC&\
7.    allowPublicKeyRetrieval=true
8.  jdbc.username=root
9.  jdbc.password=root
10.
11. # 书籍文件和封面文件存放路径
12. book_path = D:/lianshu/ebooks/
13. book_cover_path = D:/lianshu/bookCovers/
14.
15. # 后台管理系统配置
16. # 后台管理系统的路径为初始 URL 后加/backstage/
17. admin_user = root
18. admin_password = root1213
```

以上代码配置了数据库的相关连接信息、书籍和封面文件存放的地址、后台管理系统的账号和密码。下面编写 spring-mvc.xml 配置文件，如例 13-2 所示。

【例 13-2】 spring-mvc.xml

```
1.  <beans xmlns="http://www.springframework.org/schema/beans"
2.        xmlns:xsi="http://www.w3.org/2001/XMLSchema-instance"
3.        xmlns:mvc="http://www.springframework.org/schema/mvc"
4.        xmlns:context="http://www.springframework.org/schema/context"
5.        xsi:schemaLocation="http://www.springframework.org/schema/beans
6.          http://www.springframework.org/schema/beans/spring-beans.xsd
7.          http://www.springframework.org/schema/mvc
8.          http://www.springframework.org/schema/mvc/spring-mvc.xsd
9.          http://www.springframework.org/schema/context
10.       http://www.springframework.org/schema/context/spring-context.xsd">
11. <context:component-scan base-package="com.qianfeng.controller"/>
12.
13.     <!--@Controller 注解的必要条件-->
14.     <mvc:annotation-driven />
15.
16.     <!-- 静态资源访问 -->
17.     <mvc:default-servlet-handler/>
18.
19.     <!--视图解析器-->
20.     <bean id="viewResolver"
class="org.springframework.web.servlet.view.InternalResourceViewResolver">
21.         <property name="prefix" value="/WEB-INF/views/" />
22.         <property name="suffix" value=".jsp" />
23.     </bean>
24.     <!-- 使用 Spring 的 CommonsMultipartResolver,
25.          配置 MultipartResolver 用于上传文件-->
26.     <bean id="multipartResolver"
class="org.springframework.web.multipart.commons.CommonsMultipartResolver">
27.           <property name="maxUploadSize">
28.               <value>104857600</value>
```

```
29.          </property>
30.          <property name="maxInMemorySize">
31.             <value>40960</value>
32.          </property>
33.          <property name="defaultEncoding">
34.             <value>utf-8</value>
35.          </property>
36.          <property name="resolveLazily">
37.             <value>true</value>
38.          </property>
39.     </bean>
40.     <!--在超出上传文件限制时，
41. Spring MVC 会抛出“org.springframework.web.multipart.
MaxUploadSizeExceededException”的异常 -->
42.     <!-- 该异常是 Spring MVC 在检查上传的文件信息时抛出来的，
43.     而且此时还没有进入控制器的方法中 -->
44.     <bean id="exceptionResolver"
class="org.springframework.web.servlet.handler.SimpleMappingExceptionResolver">
45.        <property name="exceptionMappings">
46.           <props>
47.              <!--  在抛出“org.springframework.web.multipart.
MaxUploadSizeExceededException”异常时，
48.              自动跳转到/WEB-INF/views/error_fileupload.jsp 对应的页面   -->
49.              <prop
50.   key="org.springframework.web.multipart.MaxUploadSizeExceededException">
51.                 upload_failed
52.              </prop>
53.           </props>
54.        </property>
55.     </bean>
56. </beans>
```

在例 13-2 中，第 44～55 行代码负责处理上传文件过大时抛出的异常，当上传的文件过大时，跳转到上传失败页面。

在创建完基本的配置文件之后，创建日志配置文件。在 config 文件夹下创建 log4j.properties 文件，代码如下所示。

```
1.  #OFF,systemOut,logFile,logDailyFile,logRollingFile,logMail,logDB,ALL
2.  log4j.rootLogger =INFO,systemOut,logDailyFile
3.
4.  #输出到控制台
5.  log4j.appender.systemOut = org.apache.log4j.ConsoleAppender
6.  log4j.appender.systemOut.layout = org.apache.log4j.PatternLayout
7.  log4j.appender.systemOut.layout.ConversionPattern =
8.            [%-5p][%-22d{yyyy/MM/dd HH:mm:ssS}][%l]%n%m%n
9.  log4j.appender.systemOut.Threshold = INFO
10. log4j.appender.systemOut.ImmediateFlush = true
11. log4j.appender.systemOut.Target = System.out
12. log4j.logger.org.mybatis=DEBUG
```

以上代码指定日志的输出级别为 INFO、输出方式为控制台输出。下面编写 web.xml 文件，代码如例 13-3 所示。

【例 13-3】 web.xml

```
1.  <?xml version="1.0" encoding="UTF-8"?>
2.  <web-app version="3.0" xmlns="http://java.sun.com/xml/ns/javaee"
3.          xmlns:xsi="http://www.w3.org/2001/XMLSchema-instance"
4.          xsi:schemaLocation="http://java.sun.com/xml/ns/javaee
5.          http://java.sun.com/xml/ns/javaee/web-app_3_0.xsd">
6.
7.      <welcome-file-list>
8.          <welcome-file>index</welcome-file>
9.      </welcome-file-list>
10.
11.     <!-- Spring 配置文件的加载 -->
12.     <context-param>
13.         <param-name>contextConfigLocation</param-name>
14.         <param-value>classpath:springApplication.xml</param-value>
15.     </context-param>
16.
17.     <!--log4j 配置文件的加载-->
18.     <context-param>
19.         <param-name>log4jConfigLocation</param-name>
20.         <param-value>classpath:config/log4j.properties</param-value>
21.     </context-param>
22.
23.     <listener>
24.         <listener-class>
25.             org.springframework.web.util.Log4jConfigListener
26.         </listener-class>
27.     </listener>
28.
29.     <!-- Spring 监听器 -->
30.     <listener>
31.         <listener-class>
32.             org.springframework.web.context.ContextLoaderListener
33.         </listener-class>
34.     </listener>
35.
36.     <!--加载 Spring MVC 的配置文件 -->
37.     <servlet>
38.         <servlet-name>spring-mvc</servlet-name>
39.         <servlet-class>
40.             org.springframework.web.servlet.DispatcherServlet
41.         </servlet-class>
42.         <init-param>
43.             <param-name>contextConfigLocation</param-name>
44.             <param-value>classpath:spring-mvc.xml</param-value>
45.         </init-param>
46.         <load-on-startup>1</load-on-startup>
47.     </servlet>
48.
49.     <servlet-mapping>
50.         <servlet-name>spring-mvc</servlet-name>
51.         <url-pattern>/</url-pattern>
52.     </servlet-mapping>
```

```
53.
54.     <servlet-mapping>
55.         <servlet-name>spring-mvc</servlet-name>
56.         <url-pattern>/index</url-pattern>
57.     </servlet-mapping>
58.
59.     <!--配置过滤器，防止乱码的产生-->
60.     <filter>
61.         <filter-name>encodingFilter</filter-name>
62.         <filter-class>
63.             org.springframework.web.filter.CharacterEncodingFilter
64.         </filter-class>
65.         <init-param>
66.             <param-name>encoding</param-name>
67.             <param-value>utf-8</param-value>
68.         </init-param>
69.         <init-param>
70.             <param-name>forceEncoding</param-name>
71.             <param-value>true</param-value>
72.         </init-param>
73.     </filter>
74.
75.     <filter-mapping>
76.         <filter-name>encodingFilter</filter-name>
77.         <url-pattern>/</url-pattern>
78.     </filter-mapping>
79. </web-app>
```

在例 13-3 中，第 12～15 行代码指定了 Spring 配置文件的位置，第 18～21 行代码指定了 log4j 配置文件的位置。其他配置是 Web 中常见的配置，此处不讲解。

13.4 标题栏模块

标题栏模块是敛书网中每个页面都具备的模块。在敛书网中，标题栏模块分为 3 个部分，分别为敛书网 logo、书籍分类导航链接和登录注册按钮组。书籍分类导航链接用于分类查看书籍，“登录”按钮用于用户登录，“注册”按钮用于用户注册。本节将针对标题栏模块进行详细讲解。

13.4.1 标题栏的制作

【实战目标】

在敛书网中，标题栏贯穿于整个网站，起到导航的作用。敛书网的标题栏如图 13.34 所示。

图 13.34 敛书网的标题栏

通过对本小节的学习，读者可以制作一个标题栏。

【实现步骤】

1．编写标题栏对应的代码

在 common 文件夹下创建 loginHead.jsp 文件，其代码如下所示。

```
1.  <%@ page language="java"
2.          contentType="text/html; charset=UTF-8"
3.          pageEncoding="UTF-8"%>
4.
5.  <%--标题栏（包括左侧的 logo、导航链接和右侧的登录注册按钮组）--%>
6.  <div id="loginHead">
7.  <%--标题栏--%>
8.    <div  class="navbar navbar-default navbar-fixed-top">
9.        <div class="container">
10.          <div class="navbar-header">
11.             <button type="button" class="navbar-toggle collapsed"
12.                     data-toggle="collapse" data-target="#navbar"
13.                     aria-expanded="false" aria-controls="navbar">
14.                 <span class="sr-only">Toggle navigation</span>
15.                 <span class="icon-bar"></span>
16.                 <span class="icon-bar"></span>
17.                 <span class="icon-bar"></span>
18.             </button>
19.             <a class="navbar-brand" href="index">
20.               <img
21.       src="${pageContext.request.contextPath}/resources/images/logo.png">
22.             </a>
23.         </div>
24.         <div id="navbar" class="navbar-collapse collapse clearfix">
25.             <ul class="nav navbar-nav">
26.                 <li><a href="index">首页</a> </li>
27.                 <li><a href="bookList?bookType=经典文学">经典文学</a> </li>
28.                 <li><a href="bookList?bookType=通俗小说">通俗小说</a> </li>
29.                 <li><a href="bookList?bookType=计算机类">计算机类</a> </li>
30.                 <li><a href="bookList?bookType=杂志期刊">杂志期刊</a> </li>
31.             </ul>
32.             <%--在页面顶部的右侧显示“登录”和“注册”按钮--%>
33.             <%--登录注册按钮组用于触发登录或注册模态框--%>
34.             <div id="loginGroup" class="btn-group pull-right btn-group-sm">
35.                 <button type="button"
36.                         class="btn btn-default"
37.                         data-toggle="modal"
38.                         data-target="#loginModal">
39.                     登录
40.                 </button>
41.                 <button
42.                         type="button"
43.                         class="btn btn-default"
44.                         data-toggle="modal"
45.                         data-target="#regModal">
46.                     注册
47.                 </button>
48.             </div>
```

```
49.         </div>
50.         </div>
51.     </div>
52. </div>
```

下面详细讲解以上代码。

（1）代码的第 20～21 行的作用是添加标题栏中敛书网 logo 图片，当单击此图片时，跳转到首页。

（2）代码的第 25～31 行的作用是添加标题栏中的分类导航链接，当单击导航链接时，跳转到相应的书籍分类页面，并传递相应的分类信息。

（3）代码的第 35～47 行的作用是添加标题栏中的登录注册按钮组，当单击其中的按钮时，跳转到相应的登录或注册页面。

2．编写登录后标题栏对应的代码

对于用户登录后的标题栏，需要将其右侧的登录注册按钮组去掉，新增用户的头像和个人信息。在此使用替换页面的方式来实现用户登录后标题栏的效果。在 common 文件夹下创建一个新的标题栏页面文件，并将其命名为 userHead.jsp，编写其中的代码，其核心代码如下所示。

```
1.  <%@ page language="java" contentType="text/html; charset=UTF-8" pageEncoding=
"UTF-8"%>
2.
3.  <%--标题栏（包括左侧的 logo、导航链接和右侧的登录注册按钮组）--%>
4.  <div id="userHead" class="navbar navbar-default navbar-fixed-top hide">
5.      <div class="container">
6.          <div class="navbar-header">
7.              <button type="button" class="navbar-toggle collapsed"
8.                      data-toggle="collapse" data-target="#navbar"
9.                      aria-expanded="false" aria-controls="navbar">
10.                 <span class="sr-only">Toggle navigation</span>
11.                 <span class="icon-bar"></span>
12.                 <span class="icon-bar"></span>
13.                 <span class="icon-bar"></span>
14.             </button>
15.             <a class="navbar-brand" href="index">
16. <img src="${pageContext.request.contextPath}/resources/images/logo.png">
17.             </a>
18.         </div>
19.         <div id="navbar" class="navbar-collapse collapse clearfix">
20.             <ul class="nav navbar-nav">
21.                 <li><a href="index">首页</a> </li>
22.                 <li><a href="bookList?bookType=经典文学">经典文学</a> </li>
23.                 <li><a href="bookList?bookType=通俗小说">通俗小说</a> </li>
24.                 <li><a href="bookList?bookType=计算机类">计算机类</a> </li>
25.                 <li><a href="bookList?bookType=杂志期刊">杂志期刊</a> </li>
26.             </ul>
27.             <%--登录后用用户头像替换登录注册按钮组--%><!--用户头像信息-->
28.             <div id="loginedInfo" class="pull-right">
29.                 <a data-toggle="popover" data-placement="bottom" href="#">
30.                     <img class="img-circle" src="" alt="">
31.                 </a>
32.             </div>
```

```
33.
34.         </div>
35.     </div>
36. </div>
```

以上代码中的第 28～32 行的作用是添加用户的头像信息。其余代码与未登录时的标题栏对应的代码相同。

13.4.2 登录功能的实现

【实战目标】

在未登录页面单击标题栏中的“登录”按钮后，弹出登录对话框。用户在登录对话框中输入用户名和密码，然后单击“登录”按钮登录。用户登录后，退出登录对话框，标题栏刷新为登录后的状态。登录对话框如图 13.35 所示。

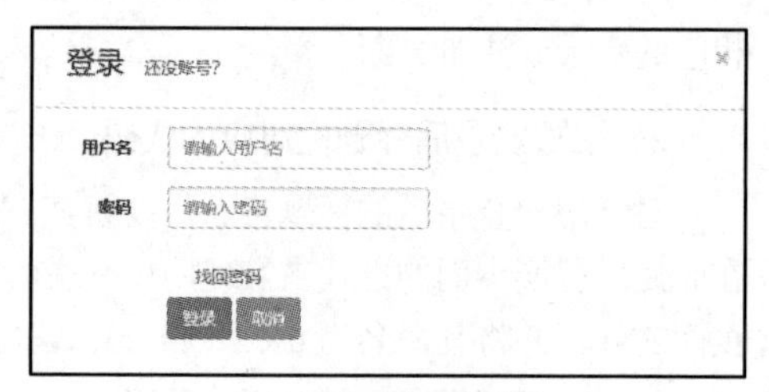

图 13.35　登录对话框

通过对此小节的学习，读者可以掌握登录对话框的制作及登录功能的实现。

【实现步骤】

1. 编写登录对话框对应的代码

在 loginHead.jsp 文件中添加如下代码。

```
1.  <div id="loginModal" class="modal fade" data-backdrop="static">
2.      <div class="modal-dialog">
3.          <div class="modal-content">
4.              <div class="modal-header">
5.                  <button type="button" class="close" data-dismiss="modal"
6.                          aria-hidden="true">
7.                      ×
8.                  </button>
9.                  <div class="container">
10.                     <span class="h3">登录</span>
11.                     <!--触发注册模态框-->
12.                     <button type="button" class="btn btn-link"
13.                             data-toggle="modal"
14.                             data-target="#regModal"
15.                             data-dismiss="modal">
16.                         还没账号?
17.                     </button>
18.                 </div>
19.             </div>
20.             <div class="modal-body">
21.                 <form id="loginForm" class="form-horizontal" role="form"
22.                       method="post"
23.                       action="login" >
24.                     <div class="form-group">
25.                         <label for="userCode"
26.                                class="control-label col-md-2">
27.                             用户名
```

```
28.                     </label>
29.                     <div class="col-md-5">
30.                         <input type="text" class="form-control"
31.                             id="userCode"
32.                             name="userCode"
33.                             placeholder="请输入用户名">
34.                     </div>
35.                 </div>
36.                 <div class="form-group">
37.                     <label for="userPassword"
38.                         class="control-label col-md-2">
39.                         密码
40.                     </label>
41.                     <div class="col-md-5">
42.                         <input type="password" class="form-control"
43.                             id="userPassword"
44.                             name="userPassword"
45.                             placeholder="请输入密码">
46.                     </div>
47.                 </div>
48.                 <div class="checkbox col-md-6 col-md-offset-2">
49.                     <a href="#" class="btn btn-link">找回密码</a>
50.                 </div>
51.                 <div class="form-group">
52.                     <div class="col-md-4 col-md-offset-2">
53.                         <button id="loginSubmit"
54.                             type="button"
55.                             class="btn btn-success">
56.                             登录
57.                         </button>
58.                         <button type="button"
59.                             class="btn btn-info"
60.                             data-dismiss="modal">
61.                             取消
62.                         </button>
63.                     </div>
64.                     <div id="errorTxt"
65.                         class="col-md-3 text-danger hide">
66.                         *用户名或密码错误
67.                     </div>
68.                 </div>
69.             </form>
70.         </div>
71.     </div>
72.   </div>
73. </div>
```

以上代码的第4～19行的作用是添加登录对话框的标题栏。以上代码的第20～70行，利用<form>表单元素构造登录对话框，当单击“登录”按钮后，用户名和密码将请求发送给控制器。

2．编写登录功能对应的代码

在controller文件夹下创建LoginController.java文件，编写登录功能对应的代码。登录功能的核心代码如下所示。

```
1.  @RequestMapping(value = "/login", method = RequestMethod.POST)
2.  @ResponseBody
3.  public Map<String, Object> index(@RequestBody User user, HttpSession session) {
4.      log.info("you are logging in!");
5.      String userCode = user.getUserCode();
6.      String password = user.getUserPassword();
7.      UserHelper userHelper = userService.getLoginUser(userCode, password);
8.      Map<String, Object> resultMap = new HashMap<String, Object>();
9.      if (null != userHelper) {
10.         session.setAttribute("userHelper", userHelper);
11.         resultMap.put("isLogined", true);
12.         resultMap.put("user", userHelper);
13.     } else {
14.         resultMap.put("isLogined", false);
15.     }
16.     return resultMap;
17. }
```

以上代码的第 7 行通过调用 UserService 类的 getLoginUser()方法来获取用户的信息。以上代码的第 10 行将用户信息封装到用户的 Session 对象中，当用户发送请求时，服务器端将获取此用户的 Session 对象，从而针对此用户进行服务。下面编写 Service 层代码。

在 service 文件夹下创建 UserService.java 文件，编写其中的代码，其核心代码如下所示。

```
1.  public UserHelper getLoginUser(String userCode, String userPassword) {
2.          User user = userDao.queryByLogin(userCode, userPassword);
3.          if (null != user) {
4.              Avatar avatar = avatarDao.queryById(user.getAvatarNum());
5.              Contribution contribution =
6.                      contributionDao.queryByValue(user.getContribution());
7.              UserHelper userHelper = new UserHelper();
8.              userHelper.setId(user.getId());
9.              userHelper.setUserCode(user.getUserCode());
10.             userHelper.setUserName(user.getUserName());
11.             userHelper.setEmail(user.getEmail());
12.             userHelper.setAvatar(avatar);
13.             userHelper.setContributionValue(user.getContribution());
14.             userHelper.setContribution(contribution);
15.             userHelper.setCreationDate(user.getCreationDate());
16.             return userHelper;
17.         } else {
18.             return null;
19.         }
20.     }
```

从以上代码可以看出，通过调用 UserDao 对象的 queryByLogin()方法，并将用户名和密码传入，就可以得到用户的信息。在此需要注意的是，User 对象中的属性有些并不适合发送给前端，因此通过 UserHelper 类来封装发送给前端的信息。以上代码的第 8～16 行设置了 UserHelper 的属性，并将其返回。下面创建 UserDao.java 文件（此次讲解作为示例，在以后的业务流程中将不再讲解 DAO 层、Mapper 层代码）。

在 dao 文件夹下创建 UserDao 接口文件，编写其中的代码，其核心代码如下所示。

```
User queryByLogin(@Param("userCode") String userCode,
                  @Param("userPassword") String userPassword);
```

编写此接口对应的 Mapper 文件。在 mapper 文件夹下创建 UserMapper.xml 文件，编写其中的代码，其核心代码如下所示。

```
1.  <?xml version="1.0" encoding="utf-8" ?>
2.  <!DOCTYPE mapper PUBLIC "-//mybatis.org//DTD Mapper 3.0//EN"
3.          "http://mybatis.org/dtd/mybatis-3-mapper.dtd" >
4.
5.  <mapper namespace="com.qianfeng.dao.UserDao">
6.
7.      <resultMap id="userMap" type="com.qianfeng.domin.User">
8.          <id column="id" property="id"/>
9.          <result column="userCode" property="userCode"/>
10.         <result column="userPassword" property="userPassword"/>
11.         <result column="userName" property="userName"/>
12.         <result column="email" property="email"/>
13.         <result column="avatarNum" property="avatarNum"/>
14.         <result column="contribution" property="contribution"/>
15.         <result column="creationDate" property="creationDate"/>
16.     </resultMap>
17.     <sql id="user_columns">
18.         id, userCode, userPassword, userName, email,
19.         avatarNum, contribution, creationDate
20.     </sql>
21.     <select id="queryByLogin" resultMap="userMap">
22.         SELECT
23.         <include refid="user_columns"/>
24.         FROM user where
25.         userCode=#{userCode} and userPassword=#{userPassword}
26.     </select>
27. </mapper>
```

在以上代码中，使用查询语句查询数据库中的数据，将查询结果封装成 User 对象并返回给前端。

3. 登录后页面的逻辑代码的编写

当用户登录后，调用登录成功方法 logininSuccess()，其代码如下所示。

```
1.  function logininSuccess(o) {
2.          $(".uploadBtn button").tooltip("destroy")
3.          console.log("登录成功! " + o.user.userCode + "," + o.user.userName + "," +
o.user.contributionValue)
4.          $("#loginHead").addClass("hide")
5.          $("#userHead").removeClass("hide")
6.          e(o), $avatar = $("#navbar #loginedInfo a img")
7.          $avatar.attr("src", o.user.avatar.avatar_img)
8.          $avatar.attr("alt", o.user.avatar.avatar_txt)
9.      }
```

以上代码的第 4～5 行表示登录后的标题栏需要进行的切换操作。然后添加个人信息框，具体代码如下所示。

```
1.  function e(e) {
2.      $("#loginHead").addClass("hide"), $("#userHead").removeClass("hide"),
3.          $("#navbar #loginedInfo a").popover({
4.          placement: "bottom",
5.          html: !0,
```

```
6.          title: '<div class="h5 text-center"><span id="userName" ' +
7.              'class="glyphicon glyphicon-user"></span></div>',
8.          content: '<div class="text-center">' +
9.              '</div>        <div class="text-success text-center">贡献值: ' +
10.             '<span id="contribution" class="badge badge-important"></span>' +
11.             '</div><button id="toPersonInfo" class="btn btn-link">' +
12.             '个人资料' +
13.             '</button>' +
14.             '<a href="#" id="exit" class="btn btn-link" >注销</a>'
15.     }).on("mouseenter", function () {
16.         var e = this;
17.         $(this).popover("show"), $(this).siblings(".popover").
18.         on("mouseleave", function () {
19.             $(e).popover("hide")
20.         })
21.     }).on("mouseleave", function () {
22.         var e = this;
23.         setTimeout(function () {
24.             $(".popover:hover").length || $(e).popover("hide")
25.         }, 100)
26.     }).on("shown.bs.popover", function () {
27.         $("#exit").on({
28.             click: function () {
29.                 setLogout();
30.                 window.location.href = "logout"
31.             }
32.         }), $("#userName").text(" " + e.user.userName),
33.             $("#contribution").text(e.user.contributionValue),
34.             $("#toPersonInfo").on({
35.             click: function () {
36.                 window.location.href = "person"
37.             }
38.         })
39.     })
40. }
```

以上代码的第 3～15 行的作用是添加个人信息框。个人信息框中包括个人贡献值、“注销”按钮和个人昵称。以上代码的第 27～32 行是单击“注销”按钮后执行的代码，当单击“注销”按钮后，调用 setLogout()方法将 cookie 注销，并跳转到 logout 页面。

13.4.3 注册功能的实现

【实战目标】

在未登录页面单击标题栏中的“注册”按钮，弹出注册对话框。用户在注册对话框中设置用户名和密码等信息后，单击“注册”按钮即可完成注册成功。注册对话框如图 13.36 所示。

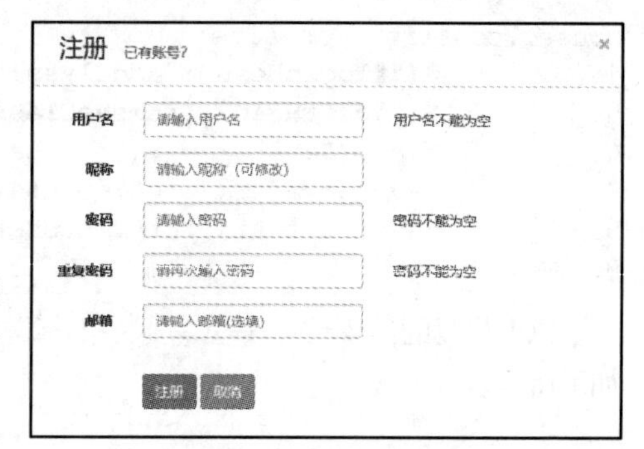

图 13.36　注册对话框

通过对此小节的学习，读者可以掌握注册对话框的制作方法及其逻辑代码的编写。

【实现步骤】

1．编写注册对话框对应的代码

注册对话框对应的代码与登录对话框对应的代码类似，在此不讲解。

2．注册功能对应代码的编写

在 controller 文件夹下创建 RegisterController.java 文件，其核心代码如下所示。

```
1.  @RequestMapping(value = "register", method = RequestMethod.POST)
2.  @ResponseBody
3.  public Map<String, Object> register(@RequestBody User user) {
4.      Map<String, Object> resultMap = new HashMap<String, Object>();
5.      userService.addUser(user);
6.      log.info("you have registered!");
7.      return resultMap;
8.  }
```

以上代码的第 5 行通过调用 UserService 类的新增方法 addUser()，将注册的用户名和密码等信息放到数据库中完成注册。

13.5 书籍展示模块

书籍展示模块是敛书网的核心模块。在敛书网中，书籍展示模块分为书籍列表、书籍分类展示页和书籍详情页。本节将详细讲解书籍展示模块各个部分对应代码的编写。

13.5.1 书籍列表的制作

【实战目标】

用户访问敛书网后，网站展示的页面是敛书网首页，首页中部的 3 个列表是书籍列表。敛书网首页如图 13.37 所示。

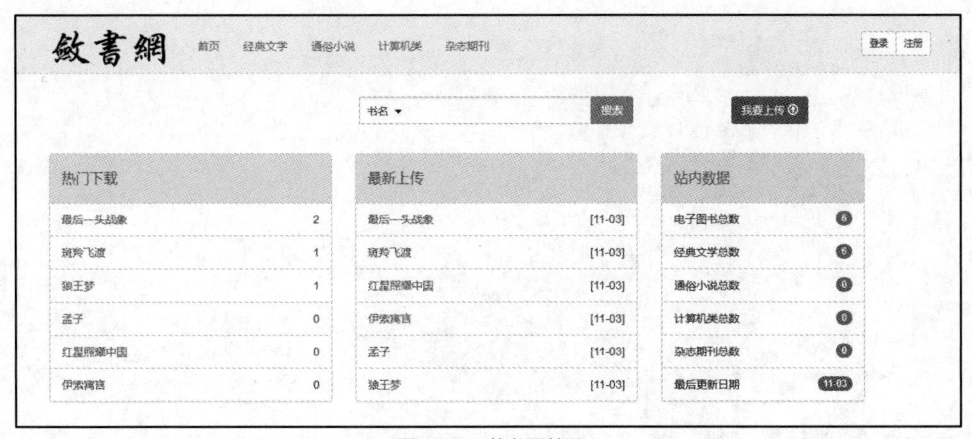

图 13.37 敛书网首页

从图 13.37 中可以看出，整个页面分为 4 个部分，分别为标题栏、搜索框、“我要上传”按钮和书籍列表。下面使用一个页面完成相应功能代码的编写。通过对本小节的学习，读者可以掌握首页和书籍列表对应代码的编写。

【实现步骤】

1. 编写首页代码

在 views 文件夹下创建 main.jsp 文件，编写首页中相关的代码。下面分步实现首页中相应的功能。

（1）实现标题栏的功能，代码如下所示。

```
<%--引入标题栏--%>
<%@include file="common/loginHead.jsp"%>
<%--引入登录后的标题栏--%>
<%@include file="common/userHead.jsp"%>
```

以上代码将 13.4.1 小节完成的标题栏引入首页，通过 JavaScript 来判断用户是否登录。如果用户已经登录，使用 userHead.jsp 文件；如果用户没有登录，则使用 loginHead.jsp 文件。

（2）实现搜索框和“我要上传”按钮的功能，在此仅做出页面，其底层实现将在 13.7 节和 13.8 节讲解，代码如下所示。

```
1.  <div class="leaderboard">
2.      <%--搜索框--%>
3.      <form class="search col-md-4 col-sm-6 col-md-offset-4"
4.            action="bookSearch"
5.            method="get">
6.          <div class="input-group">
7.              <div class="selectDiv">
8.                  <select name="searchBy" class="select">
9.                      <option class="option"
10.                             value="book_title"
11.                             selected="selected">
12.                         书名
13.                     </option>
14.                     <option class="option" value="book_author">
15.                         作者
16.                     </option>
17.                 </select>
18.             </div>
19.             <input name="searchTxt"
20.                    class="form-control"
21.                    type="text">
22.             <div class="input-group-btn">
23.                 <button class="btn btn-success"
24.                         type="submit">
25.                     搜索
26.                 </button>
27.             </div>
28.         </div>
29.     </form>
30.     <%-- “我要上传” 按钮--%>
31.     <div class="uploadBtn col-md-3 col-sm-4 col-md-offset-1">
32.         <button class="btn btn-primary" type="button">
33.             我要上传 
```

```
34.             <span class="glyphicon glyphicon-upload"></span>
35.         </button>
36.     </div>
37. </div>
```

在以上代码中，使用<select>、<input>和<button>等标签构造搜索框和“我要上传”按钮。

（3）实现书籍列表的功能。首先实现书籍列表中的热门下载部分的功能，代码如下所示。

```
1.  <div class="panel panel-info">
2.      <div class="panel-heading"><h4>热门下载</h4></div>
3.      <ul class="list-group">
4.          <c:forEach items="${rankingBooks1}" var="rankingBook">
5.              <li class="list-group-item">
6.                  <a href="bookDetail?bookID=${rankingBook.id}">
7.                          ${rankingBook.bookName}
8.                  </a>
9.                  <span class="pull-right">
10.                         ${rankingBook.download_times}
11.                 </span>
12.             </li>
13.         </c:forEach>
14.     </ul>
15. </div>
```

在以上代码中，接收控制器传递的热门书籍信息，并通过JSP将其渲染到页面上。以上代码的第6～8行使用<a>标签实现书籍名称的显示，当单击书籍名称时，跳转到相应的书籍详情页。

接下来实现书籍列表中的最新上传部分的功能，代码如下所示。

```
1.  <div class="panel panel-success">
2.      <div class="panel-heading"><h4>最新上传</h4></div>
3.      <ul class="list-group">
4.          <c:forEach items="${rankingBooks}" var="rankingBook">
5.              <li class="list-group-item">
6.                  <a href="bookDetail?bookID=${rankingBook.id}">
7.                      ${rankingBook.bookName}
8.                  </a>
9.                  <span class="pull-right">[${rankingBook.uploadDate}]</span>
10.             </li>
11.         </c:forEach>
12.     </ul>
13. </div>
```

最新上传部分与热门下载部分的代码相似，在此不讲解。最后，完成站内数据部分的功能，代码如下所示。

```
1.  <div class="panel panel-danger">
2.    <div class="panel-heading"><h4>站内数据</h4></div>
3.    <ul class="list-group">
4.      <li class="list-group-item">电子图书总数
5.          <span class="badge pull-right">${sumOfBooks}</span> </li>
6.      <li class="list-group-item">经典文学总数
7.          <span class="badge pull-right">${sumOfTypeBooks.get(0)}</span> </li>
8.      <li class="list-group-item">通俗小说总数
9.          <span class="badge pull-right">${sumOfTypeBooks.get(1)}</span> </li>
```

```
10.     <li class="list-group-item">计算机类总数
11.         <span class="badge pull-right">${sumOfTypeBooks.get(2)}</span> </li>
12.     <li class="list-group-item">杂志期刊总数
13.         <span class="badge pull-right">${sumOfTypeBooks.get(3)}</span> </li>
14.     <li class="list-group-item">最后更新日期
15.         <span class="badge pull-right">${maxUploadDate}</span> </li>
16.   </ul>
17. </div>
```

从以上代码中可以看出，站内数据部分的代码主要由<span>标签组成，其中的数据从 sumOfTypeBooks 数组中获取。下面编写后台业务逻辑代码。

2．编写书籍列表的业务逻辑代码

在 controller 文件夹下创建 MainController.java 文件，编写其中的代码，其核心代码如下所示。

```
1.  @RequestMapping(value = "/index")
2.  public String index(Model model) {
3.      /*站内数据部分所需的数据*/
4.          /*电子图书的总数*/
5.          int sumOfBooks = bookService.queryNumberOfBooks();
6.          model.addAttribute("sumOfBooks", sumOfBooks);
7.          /*不同类型书籍的数目*/
8.          List<Integer> everyTypeBooks;
9.          everyTypeBooks = bookService.queryNumberOfSomeTypeBooks();
10.         model.addAttribute("sumOfTypeBooks", everyTypeBooks);
11.         /*最后上传图书的日期*/
12.         String maxUploadDate = bookService.getMaxUploadDate();
13.         model.addAttribute("maxUploadDate", maxUploadDate);
14.
15.         /*最新上传部分所需的数据*/
16.         List<RankingBook> rankingBooks;
17.         rankingBooks = bookService.queryByUploadedDate();
18.         model.addAttribute("rankingBooks", rankingBooks);
19.         /*热门下载部分所需的数据*/
20.         List<RankingBook> rankingBooks1;
21.         rankingBooks1 = bookService.queryByDownloadTimes();
22.         model.addAttribute("rankingBooks1", rankingBooks1);
23.
24.         return "main";
25. }
```

从以上代码中可以看出，当访问 index 页面时，控制器通过调用 BookService 类的方法，查询出前端需要的数据，然后将其放入 Model 中并返回给前端。下面编写 BookService 类的代码。

在 service 文件夹下创建 BookService.java 文件，编写其中的代码，其核心代码如下所示。

```
1.  public int queryNumberOfBooks() {
2.      return bookDao.queryNumberOfBooks();
3.  }
4.
5.  public List<Integer> queryNumberOfSomeTypeBooks() {
6.      List<Integer> result = new ArrayList<Integer>();
7.      for (int i = 1; i < 5; i++) {
8.          int sum = bookDao.queryNumberOfSomeTypeBooks(i);
```

```
9.         result.add(sum);
10.     }
11.     return result;
12. }
13.
14. public String getMaxUploadDate() {
15.     Date date = uploadDao.getMaxUploadDate();
16.     SimpleDateFormat dateFormat = new SimpleDateFormat("MM-dd");
17.     String maxUploadDate = "";
18.     if (null != date) {
19.         maxUploadDate = dateFormat.format(date);
20.     }
21.     return maxUploadDate;
22. }
23.
24. public List<RankingBook> queryByUploadedDate() {
25.     List<RankingBook> rankingBooks = new ArrayList<RankingBook>();
26.     List<Upload> uploadRecords;
27.     uploadRecords = uploadDao.queryByUploadedDate();
28.     Book book;
29.     SimpleDateFormat dateFormat = new SimpleDateFormat("MM-dd");
30.     for (Upload upload : uploadRecords) {
31.         RankingBook rankingBook = new RankingBook();
32.         rankingBook.setId(upload.getUploadedBook());
33.         book = bookDao.queryById(upload.getUploadedBook());
34.         rankingBook.setBookName(book.getBook_title());
35.         String uploadDate = dateFormat.format(upload.getUploadedDate());
36.         rankingBook.setUploadDate(uploadDate);
37.         rankingBooks.add(rankingBook);
38.     }
39.     return rankingBooks;
40. }
41.
42. public List<RankingBook> queryByDownloadTimes() {
43.     List<RankingBook> rankingBooks = new ArrayList<RankingBook>();
44.     List<Book> books;
45.     books = bookDao.queryByDownloadTimes();
46.     for (Book book : books) {
47.         RankingBook rankingBook = new RankingBook();
48.         rankingBook.setId(book.getId());
49.         rankingBook.setBookName(book.getBook_title());
50.         rankingBook.setDownload_times(book.getDownload_times());
51.         rankingBooks.add(rankingBook);
52.     }
53.     return rankingBooks;
54. }
```

下面详细讲解以上代码中的5个方法。

（1）queryNumberOfBooks()方法的作用是查询出电子图书的总数，并将其返回给前端。

（2）queryNumberOfSomeTypeBooks()方法的作用是查询出每种类别的图书数量，并将其封装成数组返回给前端。

（3）getMaxUploadDate()方法的作用是查询出最后一次上传图书的时间，并将其返回给前端。

（4）queryByUploadedDate()方法的作用是查询出最近上传的6本图书，并将其返回给前端。

（5）queryByDownloadTimes()方法的作用是查询出下载次数最多的6本图书，并将其返回给前端。

因内容过多，DAO 层代码与 Mapper 层代码不在此介绍，具体代码分别见项目源码包中 dao 文件夹下的 BookDao.java 文件和 mapper 文件夹下的 BookMapper.xml 文件。

13.5.2 书籍分类展示页的制作

【实战目标】

在敛书网中，用户通过单击标题栏中的分类导航链接分类浏览书籍，书籍分类展示页如图 13.38 所示。

图 13.38　书籍分类展示页

从图 13.38 中可以看出，整个页面分为 4 个部分，第一部分为标题栏，第二部分为左侧的分类导航列表，第三部分为搜索框和“我要上传”按钮，第四部分为分类书籍列表。下面使用一个页面实现各个部分的功能。通过对此小节的学习，读者可以制作书籍分类展示页。

【实现步骤】

1．编写书籍分类展示页的代码

在 views 文件夹下创建 bookList.jsp 文件。下面分步实现书籍分类展示页。

（1）实现标题栏的功能，此处与 13.4.1 小节的标题栏相同，其代码在此不讲解。

（2）实现左侧的分类导航列表的功能。在此需要注意的是，此处需要展示分类书籍列表，所以在标题栏中单击分类导航链接时，应该携带书籍分类信息访问相应的控制器，随后在数据库中查找相应的书籍列表信息，并跳转到相应的页面。编写分类导航列表的代码，代码如下所示。

```
1.  <div class="categories col-md-2 col-sm-6 col-xs-12">
2.      <ul class="nav nav-pills nav-stacked">
3.          <c:forEach items="${smallTypesOfBook}" var="smallType">
4.              <li>
5.                  <a href="bookList?bookType=${bookType}&smallType=$
{smallType.small_type_name}">
6.                      ${smallType.small_type_name}
7.                  </a>
8.              </li>
9.          </c:forEach>
10.     </ul>
```

```
11. </div>
```

以上代码的第 3 行从 Model 中获取相应的书籍类别，并将其渲染到左侧的分类导航列表中。左侧的分类导航列表是一个<a>标签，当此标签被单击时，分类书籍列表的内容更新。

（3）实现搜索框和“我要上传”按钮的功能。此处与 13.5.1 小节中的搜索框和“我要上传”按钮相同，其代码在此不讲解。

（4）编写分类书籍列表的代码，分类书籍列表分为上侧的分类栏、中间的书籍展示栏和下侧的分页栏 3 个部分。上侧的分类栏和中间的书籍展示栏的代码如下所示。

```
1.  <div class="panel panel-info">
2.      <%--分类栏--%>
3.      <div class="panel-heading">
4.          <ul class="breadcrumb">
5.              <li><a href="index">首页</a> <span class="divider"></span></li>
6.              <li>
7.                  <a id="navCurrent1"
8.                    href="bookList?bookType=${bookType}">${bookType}
9.                  </a>
10.                 <span class="divider"></span>
11.             </li>
12.             <c:if test="${smallType != null}">
13.                 <li>
14.                     <a id="navCurrent2"
15.               href="bookList?bookType=${bookType}&smallType=${smallType}">
16.                         ${smallType}
17.                     </a>
18.                     <span class="divider"></span>
19.                 </li>
20.             </c:if>
21.         </ul>
22.     </div>
23.     <%--书籍展示栏--%>
24.     <div class="panel-body">
25.
26.         <c:forEach items="${books}" var="book">
27.             <div class="col-md-3 col-sm-6 col-xs-12">
28.                 <div class="thumbnail">
29.                     <img src="getBookCover?coverPath=${book.book_cover}"
30.                          alt="${book.book_title}">
31.                     <div class="caption">
32.                         <div class="bookTitle text-center text-success"
33.                                title="${book.book_title}">
34.                                 ${book.book_title}
35.                         </div>
36.                         <div class="btn-block text-center">
37.                             <a
38.        href="book_download?bookID=${book.id}&filePath=${book.book_file}"
39.                                 class="btn btn-link">
40.                                 下载
41.                             </a>
42.                 <a href="bookDetail?bookID=${book.id}" class="btn btn-link">
43.                     详情
44.                 </a>
```

```
45.                         </div>
46.                     </div>
47.                 </div>
48.             </div>
49.         </c:forEach>
50.
51.     </div>
52.
53. </div>
```

以上代码中的第 5 行的作用是添加分类栏的首页链接。代码的第 6～11 行的作用是添加一级分类标题，一级分类标题永远存在，所以不必判断它是否为空。

以上代码的第 12～20 行的作用是添加二级分类标题，如果二级分类标题为空，则表示此页面是通过一级标题链接进入的；如果不为空，则表示此页面是通过二级标题链接进入的。当二级标题不为空时，将二级分类标题加入页面中。

以上代码的第 26～49 行的作用是添加书籍栏，书籍栏的内容是一个集合，通过<c:forEach>标签即可将集合中的元素取出，并通过<div>标签展示在页面中。

分页栏的相关内容不要求掌握，详细代码请见项目源码包中的 bookList.jsp 文件。

2. 编写书籍分类展示页的业务代码

在 controller 文件夹下创建 BookController.java 文件，编写其中的代码，其核心代码如下所示。

```
1.  @Slf4j
2.  @Controller
3.  public class BookController {
4.
5.      @Autowired
6.      private BookService bookService;
7.
8.      @RequestMapping(value = "/bookList")
9.      public String getBookList(
10.             String bookType,
11.             String smallType,
12.             @RequestParam(value = "pageId", defaultValue = "1") int pageId,
13.             Model model) {
14.         log.info("you are visiting the books list page!");
15.         List<BookType> smallTypes;
16.         smallTypes = bookService.getSmallTypesOfBook(bookType);
17.         model.addAttribute("smallTypesOfBook", smallTypes);
18.         model.addAttribute("bookType", bookType);
19.         PageHelper page = new PageHelper();
20.         page.setCurrentPage(pageId);
21.         if (null == smallType) {
22.             int sumOfBooks = bookService.getTotalOfLTBooks(smallTypes);
23.             page.setTotalRows(sumOfBooks);
24.             List<Book> books = bookService.getLargeTypeBooks(
25.                             smallTypes, page);
26.             model.addAttribute("currentPage", pageId);
27.             model.addAttribute("totalPage", page.getTotalPage());
28.             model.addAttribute("books", books);
29.         } else {
30.             int type_id = 0;
31.             for (BookType sBookType : smallTypes) {
```

```
32.             if (sBookType.getSmall_type_name().equals(smallType)) {
33.                type_id = sBookType.getId();
34.                break;
35.             }
36.          }
37.          int sumOfBooks = bookService.getTotalOfSTBooks(type_id);
38.          page.setTotalRows(sumOfBooks);
39.          List<Book> books = bookService.getSmallTypeBooks(type_id, page);
40.          model.addAttribute("currentPage", pageId);
41.          model.addAttribute("totalPage", page.getTotalPage());
42.          model.addAttribute("books", books);
43.          model.addAttribute("smallType", smallType);
44.       }
45.       return "bookList";
46.    }
47. }
```

以上代码的第 1 行的作用是激活此类的日志，当日志被激活后，在此类中可以直接通过 log 属性输出相应的日志信息，例如第 14 行的代码，使用 log.info()来输出日志信息。

以上代码的第 15～20 行用于查找某个类型的书籍。代码的第 21 行用于判断此次获取书籍的操作是否需要限定 smallType（小类型），如果没有小类型，代码的第 22～28 行将查询到的信息封装；如果存在小类型，则代码的第 30～43 行对查询到的书籍进行筛选，最终选出符合类型和细化类型的书籍。

完成控制器代码后，编写 Service 层代码，在 service 文件夹下创建 BookService 类，代码如下所示。

```
1.  @Slf4j
2.  @Service
3.  public class BookService {
4.
5.     @Autowired
6.     private BookTypeDao bookTypeDao;
7.     @Autowired
8.     private BookDao bookDao;
9.     @Autowired
10.    private UploadDao uploadDao;
11.
12.    public List<BookType> getSmallTypesOfBook(String largeTypeName) {
13.       List<BookType> bookTypes;
14.       bookTypes = bookTypeDao.queryByLargeTypeName(largeTypeName);
15.       return bookTypes;
16.    }
17.
18.    public int getTotalOfLTBooks(List<BookType> bookTypes) {
19.       List<Integer> bookTypeIdList = new ArrayList<Integer>();
20.       for (BookType bookType : bookTypes) {
21.          bookTypeIdList.add(bookType.getId());
22.       }
23.       return bookDao.getTotalOfLTBooks(bookTypeIdList);
24.    }
25.
26.    public List<Book> getLargeTypeBooks(
27.                      List<BookType> bookTypes,
28.                      PageHelper page)
29.    {
30.       List<Integer> bookTypeIdList = new ArrayList<Integer>();
```

```
31.        for (BookType bookType : bookTypes) {
32.            bookTypeIdList.add(bookType.getId());
33.        }
34.        List<Book> books = bookDao.getLargeTypeBooks(
35.                bookTypeIdList,
36.                page.getStartRow(),
37.                page.getPageSize()
38.        );
39.        return books;
40.    }
41.
42.    public int getTotalOfSTBooks(int type_id) {
43.        return bookDao.getTotalOfSTBooks(type_id);
44.    }
45.
46.    public List<Book> getSmallTypeBooks(int type_id, PageHelper page) {
47.        List<Book> books = bookDao.getSmallTypeBooks(
48.                type_id,
49.                page.getStartRow(),
50.                page.getPageSize()
51.        );
52.        return books;
53.    }
54. }
```

在以上代码中，使用 BookDao 类和 BookTypeDao 类来完成对书籍和书籍类型的更改。

13.5.3 书籍详情页的制作

【实战目标】

单击详情链接即可进入书籍详情页，此页面包含图书封面图、图书信息和图书简介等。某一书籍详情页如图 13.39 所示。

图 13.39　书籍详情页

从图 13.39 中可以看出，整个书籍详情页分为标题栏和图书详情两个部分。本小节将详细讲解书籍详情页的制作方法。

【实现步骤】

1．编写书籍详情页的代码

在 views 文件夹下创建 bookDetail.jsp 文件，因为此处的标题栏代码与 13.4.1 小节中的相同，所以直接编写图书详情部分的代码，其核心代码如下所示。

```
1.  <div class="panel panel-danger">
2.      <div class="panel-heading">
3.          <span class="text-primary">图书详情</span>
4.      </div>
5.      <div class="panel-body">
6.  
7.          <div class="book-cover col-md-4">
8.              <img src="getBookCover?coverPath=${book.book_cover}"
9.                  alt="${book.book_title}">
10.         </div>
11. 
12.         <div class="book-labels col-md-5 clearfix">
13.             <p>
14.                 <label class="label-name">图书名称：</label>
15.                 <label class="label-value">${book.book_title}</label>
16.             </p>
17.             <p>
18.                 <label class="label-name">作者：</label>
19.                 <label class="label-value">${book.book_author}</label>
20.             </p>
21.             <p>
22.                 <label class="label-name">出版时间：</label>
23.                 <label class="label-value"><fmt:formatDate
24.                         value="${book.book_pubYear}"
25.                         pattern="yyyy-MM"/></label>
26.             </p>
27.             <p>
28.                 <label class="label-name">上传者：</label>
29.                 <label class="label-value">${uploader}</label>
30.             </p>
31.             <p>
32.                 <label class="label-name">上传时间：</label>
33.                 <label class="label-value">${uploadedDate}</label>
34.             </p>
35.             <p>
36.                 <label class="label-name">下载：</label>
37.                 <label class="label-value">
38.                     <a
39.     href="book_download?bookID=${book.id}&filePath=${book.book_file}"
40.                         class="btn btn-sm btn-info">
41.                         ${format}
42.                         <i class="glyphicon glyphicon-download-alt"></i>
43.                     </a>
44.                 </label>
45.             </p>
46.             <p>
47.                 <label class="label-name">简介：</label>
48.                 <label class="label-value">${book.book_summary}</label>
49.             </p>
50.         </div>
51. 
52.         <div class="col-md-3">
```

```
53.            <div class="alert alert-warning" role="alert">
54.                <p>本站提供的所有资源仅供学习和研究使用，请勿用于商业用途。谢谢合作。</p>
55.            </div>
56.        </div>
57.    </div>
58. </div>
```

在以上代码中，使用<div>和<label>标签搭建页面的框架，使用<a>标签创建下载按钮。页面创建完成后，使用“${}”来接收后端的信息，并将其渲染到页面中。

2．编写书籍详情页的业务代码

前端需要接收的参数有图片资源和 Book 类中的属性。按照需求在 BookController.java 文件中编写代码，其核心代码如下所示。

```
1.  @RequestMapping(value = "/bookDetail")
2.  public String bookDetail(long bookID, Model model) {
3.      Book book;
4.      book = bookService.getBookDetail(bookID);
5.      Upload upload;
6.      upload = bookService.getUploadInfo(bookID);
7.      Date uploadedDate = upload.getUploadedDate();
8.      SimpleDateFormat dateFormat = new SimpleDateFormat("yyyy-MM-dd");
9.      String uploadDate = dateFormat.format(uploadedDate);
10.     User user;
11.     user = userService.queryById(upload.getUploader());
12.     model.addAttribute("book", book);
13.     model.addAttribute("uploadedDate", uploadDate);
14.     model.addAttribute("uploader", user.getUserName());
15.     model.addAttribute("format", book.getBook_format().toUpperCase());
16.     log.info("you are looking up the book:" + book.getBook_title());
17.     return "bookDetail";
18. }
19.
20. @RequestMapping(value = "/getBookCover")
21. public void getBookCover(String coverPath, HttpServletResponse response) {
22.     InputStream in = null;
23.     BufferedInputStream bis = null;
24.     OutputStream out = null;
25.     BufferedOutputStream bos = null;
26.     File file = new File(coverPath);
27.     if (!file.exists() || file.isDirectory()) {
28.         return;
29.     }
30.     try {
31.         in = new FileInputStream(coverPath);
32.         bis = new BufferedInputStream(in);
33.         byte[] data = new byte[1024];
34.         int bytes = 0;
35.         out = response.getOutputStream();
36.         bos = new BufferedOutputStream(out);
37.         while ((bytes = bis.read(data, 0, data.length)) != -1) {
38.             bos.write(data, 0, bytes);
39.         }
40.         bos.flush();
41.     } catch (IOException e) {
```

```
42.         e.printStackTrace();
43.     } finally {
44.         try {
45.             if (bos != null) {
46.                 bos.close();
47.             }
48.             if (out != null) {
49.                 out.close();
50.             }
51.             if (bis != null) {
52.                 bis.close();
53.             }
54.             if (in != null) {
55.                 in.close();
56.             }
57.         } catch (IOException e) {
58.             e.printStackTrace();
59.         }
60.     }
61. }
```

下面详细讲解以上代码。

代码的第 4 行根据书籍的 bookID 获取对应书籍的信息；代码的第 6 行根据书籍的 bookID 获取对应的上传信息；代码的第 11 行根据上传者来获取对应的用户信息；代码的第 12～15 行将查询到的信息放入 Model 中，并将其传到前端。因为 Service 层的查询方法较简单，在此不做讲解。

在前端代码中，通过 src 调用代码第 21 行的 getBookCover()方法，从而获取图片的文件流。代码的第 31～49 行根据文件的地址将图片的文件流输出，并在前端页面展示。

13.6 书籍搜索模块

书籍搜索模块是用户经常用到的功能模块，它没有过多的前端代码，而且其业务代码相比于其他模块较简单。本节将简单介绍书籍搜索页代码的编写。

书籍搜索页的制作

【实战目标】

在首页的搜索框中输入图书的名称，单击“搜索”按钮，跳转到书籍搜索页。书籍搜索页如图 13.40 所示。

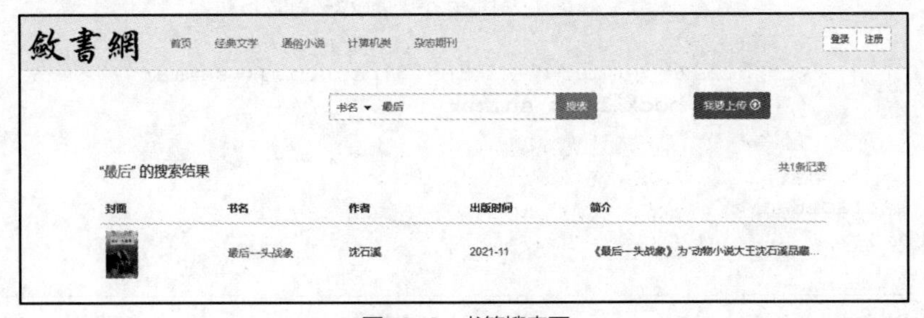

图 13.40 书籍搜索页

从图 13.40 中可以看出，书籍搜索页分为 4 个部分，分别为标题栏、搜索框、“我要上传”按钮和书籍列表。因为标题栏、搜索框和“我要上传”按钮已经在 13.5.1 小节中讲解完毕，所以本节将讲解书籍列表代码的编写。

【实现步骤】

1. 编写书籍列表的代码

在 views 文件夹下创建 searchResult.jsp 文件，编写其中的代码，其核心代码如下所示。

```
1.  <table class="table table-hover">
2.      <caption class="h4">
3.          "<span class="text-primary">${searchTxt}</span>" 的搜索结果
4.          <span class="pull-right small">共${books.size()}条记录</span>
5.      </caption>
6.      <thead>
7.          <tr>
8.              <th class="col-md-2 col-xs-6">封面</th>
9.              <th class="col-md-2 col-xs-6">书名</th>
10.             <th class="col-md-2 col-xs-6">作者</th>
11.             <th class="col-md-2 col-xs-6">出版时间</th>
12.             <th class="col-md-4 col-xs-12">简介</th>
13.         </tr>
14.     </thead>
15.     <tbody>
16.         <c:forEach items="${books}" var="book">
17.             <tr>
18.                 <td class="bookCover">
19.                     <img src="getBookCover?coverPath=${book.book_cover}"
20.                          alt="${book.book_title}">
21.                 </td>
22.                 <td class="bookInfo">
23.                     <a href="bookDetail?bookID=${book.id}"
24.                        class="btn btn-link"
25.                        title="${book.book_title}">
26.                            ${book.book_title}
27.                     </a>
28.                 </td>
29.                 <td class="bookInfo">
30.                         ${book.book_author}
31.                 </td>
32.                 <td class="bookInfo">
33.                     <fmt:formatDate value="${book.book_pubYear}"
34.                                     pattern="yyyy-MM"/>
35.                 </td>
36.                 <td class="summary" title="${book.book_summary}">
37.                         ${book.book_summary}
38.                 </td>
39.             </tr>
40.         </c:forEach>
41.     </tbody>
42. </table>
```

下面详细讲解以上代码。

整个书籍列表由一个<table>表格组成，在代码的第 16～40 行使用<c:forEach>标签循环构造书籍列表，同时在每一次的循环中获取书籍信息，并将其渲染到页面中。

此页面需要的数据主要有搜索内容和书籍信息。搜索内容通过以上代码的第 3 行进行获取，书籍信息通过以上代码的第 16～40 行进行获取。

2．编写书籍列表的业务逻辑代码

在 BookController.java 文件中添加书籍搜索方法，其代码如下所示。

```
1.  @RequestMapping(value = "/bookSearch")
2.  public String bookSearch(String searchBy, String searchTxt,
3.                           Model model) throws ParseException {
4.      log.info("you are searching book!");
5.      log.info("The search context is " + searchTxt);
6.      List<Book> books = bookService.searchBook(searchBy, searchTxt);
7.      model.addAttribute("books", books);
8.      model.addAttribute("searchTxt", searchTxt);
9.      return "searchResult";
10. }
```

在以上代码中，通过 BookService 类的 searchBook()方法来搜索符合条件的书籍。下面编写 Service 层的业务代码。

在 service 文件夹下的 BookService.java 文件中添加 searchBook()方法，详细代码如下所示。

```
1.  public List<Book> searchBook(String searchBy,
2.                               String searchTxt) throws ParseException {
3.      List<Book> books;
4.      if (searchBy.equals("book_title")) {
5.          books = bookDao.searchBookByTitle(searchTxt);
6.      } else {
7.          books = bookDao.searchBookByAuthor(searchTxt);
8.      }
9.      return books;
10. }
```

以上代码的第 4～8 行是主要的业务逻辑代码，此段代码首先判断搜索的种类，然后根据搜索种类调用相应的方法查询书籍。此处的搜索采用模糊查询，模糊查询的 XML 文件代码如下所示。

```
1.  <!--通过书名模糊查询书籍-->
2.  <select id="searchBookByTitle" parameterType="String" resultMap="bookMap">
3.      select
4.      <include refid="book_columns"/>
5.      from book
6.      where book_title LIKE concat('%', #{searchTxt}, '%')
7.      order by download_times desc
8.      limit 10
9.  </select>
10. <!--通过作者名模糊查询书籍-->
11. <select id="searchBookByAuthor"
12.             parameterType="String"
13.             resultMap="bookMap">
14.     select
15.     <include refid="book_columns"/>
16.     from book
17.     where book_author LIKE concat('%', #{searchTxt}, '%')
```

```
18.     order by download_times desc
19.     limit 10
20. </select>
```

以上代码的第 3～8 行根据书名模糊查询书籍；以上代码的第 14～19 行根据作者名模糊查询书籍。

13.7 书籍上传模块

书籍上传模块需要用户登录之后才可以使用，此模块没有过多的页面，但其后端业务逻辑十分复杂，需要读者深入理解。本节将详细讲解书籍上传模块的代码编写。

书籍上传页的制作

【实战目标】

单击“我要上传”按钮，当用户处于登录状态时，跳转到书籍上传页。书籍上传页如图 13.41 所示。

图 13.41　书籍上传页

从图 13.41 中可以看出，书籍上传页分为两个部分，分别是标题栏和书籍上传框。标题栏在 13.4.1 小节中讲解完毕，因此本小节将主要讲解书籍上传框的代码编写。

【实现步骤】

1. 编写书籍上传框的代码

在 views 文件夹下创建 upload.jsp 文件，编写其中的代码，忽略普通的输入框组件，其核心代码如下所示。

```
1.  <div class="form-group">
2.      <label for="fileUpload"
3.             class="control-label col-md-1 text-success">
4.          文件
5.      </label>
6.      <div class="input-group col-md-5">
```

```
7.          <input id="fileInfo"
8.                class="form-control"
9.                readonly type="text"
10.               placeholder="支持TXT、EPUb、MOBI和PDF格式">
11.         <span class="input-group-addon btn btn-success btn-file">
12.            Browse <input id="fileUpload" name="bookFile"
13.                          type="file">
14.         </span>
15.      </div>
16. </div>
17. <div class="form-group">
18.      <label for="imageUpload"
19.            class="control-label col-md-1 text-success">
20.         封面
21.      </label>
22.      <div class="input-group col-md-5">
23.         <input id="imageInfo" class="form-control" readonly
24.               type="text"
25.               placeholder="支持JPG和PNG图片格式">
26.         <span class="input-group-addon btn btn-success btn-file">
27.            Browse <input id="imageUpload"
28.                       name="bookCover"
29.                       type="file">
30.         </span>
31.      </div>
32. </div>
```

从以上代码中可以看出，将<input>的type属性设置为file即可上传文件。单击“提交”按钮，表单会被提交，表单的代码如下所示。

```
<form id="uploadForm" class="form-horizontal" action="doUpload"...>
```

从以上代码中可以看出，表单的内容被提交到doUpload控制器，下面编写控制器代码。

2. 编写书籍上传的业务代码

在controller文件夹中建立UploadController.java文件，编写其中的代码，其核心代码如下所示。

```
1.  @RequestMapping(value = "doUpload", method = RequestMethod.POST)
2.  public String doUpload(@ModelAttribute BookHelper bookHelper,
3.                    Model model, HttpSession session)
4.          throws IllegalStateException, IOException, ParseException {
5.      log.info("you are uploading a book! ");
6.      log.info("This book is " + bookHelper.getTitle() + "!");
7.      String fileName = bookHelper.getBookFile().getOriginalFilename();
8.      String bookCover = bookHelper.getBookCover().getOriginalFilename();
9.      MultipartFile bookFile = bookHelper.getBookFile();
10.     MultipartFile coverFile = bookHelper.getBookCover();
11.     if (bookFile.isEmpty()) {
12.        log.info("Uploading failed! The book you are uploading is empty!");
13.        return "upload_failed";
14.     } else if (coverFile.isEmpty()) {
15.        log.info("Uploading failed! The book cover you are uploading is empty!");
16.        return "upload_failed";
17.     } else {
18.        String typeId = "" + bookHelper.getLargeType() +
```

```
19.                         bookHelper.getSmallType();
20.         int type_id = Integer.parseInt(typeId);
21.         String format = fileName.substring(fileName.lastIndexOf('.') + 1);
22.         List<String> typeNames;
23.         typeNames = bookService.getTypeNames(type_id);
24.         String filePath_pre = (String) environment.getProperty("book_path");
25.         String filePath = filePath_pre + typeNames.get(0) +
26.                 "/" + typeNames.get(1) + "/" +
27.                 bookHelper.getTitle() + "." + format;
28.         File localBookFile = new File(filePath);
29.         if (localBookFile.exists()) {
30.             log.info("Uploading failed! The book is existed!");
31.             return "upload_failed2";
32.         }
33.         bookFile.transferTo(localBookFile);
34.         String coverPath_pre =
35.                     (String) environment.getProperty("book_cover_path");
36.         String coverPath = coverPath_pre + typeNames.get(0) +
37.                 "/" + typeNames.get(1) + "/" +
38.                 bookHelper.getTitle() + ".jpg";
39.         File localCoverFile = new File(coverPath);
40.         coverFile.transferTo(localCoverFile);
41.         log.info("The book has uploaded to local path successfully!");
42.
43.         Book book = new Book();
44.         book.setBook_title(bookHelper.getTitle());
45.         book.setBook_author(bookHelper.getAuthor());
46.         SimpleDateFormat dateFormat = new SimpleDateFormat("yyyy-MM");
47.         Date date = dateFormat.parse(bookHelper.getPubYear());
48.         book.setBook_pubYear(date);
49.         book.setBook_summary(bookHelper.getSummary());
50.         book.setType_id(type_id);
51.         book.setBook_format(format);
52.         book.setDownload_times(0);
53.         book.setBook_file(filePath);
54.         book.setBook_cover(coverPath);
55.         dateFormat = new SimpleDateFormat("yyMMdd", Locale.CHINESE);
56.         String pubDate = dateFormat.format(date);
57.         String upDate = dateFormat.format(new Date());
58.         int random = new Random().nextInt(900) + 100;
59.         String idStr = "" + typeId + pubDate + upDate + random;
60.         long bookID = Long.parseLong(idStr);
61.         log.info("The book id you uploaded is " + bookID);
62.         book.setId(bookID);
63.         bookService.uploadBook(book);
64.         UserHelper userHelper =
65.                     (UserHelper) session.getAttribute("userHelper");
66.         bookService.updateRecords(userHelper.getId(), bookID);
67.         userService.updateUserContribution(2, userHelper.getId());
68.         model.addAttribute("bookID", bookID);
69.         log.info("you are coming to the uploading successful page!");
70.         return "upload_success";
71.     }
72. }
```

下面详细讲解以上代码。

（1）前端将表单提交的信息传入控制器，代码的第 2 行使用@ModelAttribute 注解将参数封装到 BookHelper 中。

（2）代码的第 7～10 行获取 BookHelper 中的数据。

（3）代码的第 20 行拼接书籍类型编号，随后在代码的 23 行通过它查询出书籍的类型名。

（4）代码的第 28 行通过路径与类型名的拼接，创建文件路径。

（5）代码的第 33 行将接收的文件传输到第 28 行代码创建的文件中。

（6）代码的第 34～38 行拼接封面图的路径，第 39、40 行代码负责将封面图传输到相应路径。

（7）代码的第 43～62 行实例化 Book 类，并设置其属性；在代码的第 63 行通过 BookService 类的 uploadBook()方法将书籍加入数据库中。

（8）代码的第 66～67 行更新记录和用户的贡献信息。

下面编写 Service 层的方法。在 BookService 类中添加 uploadBook()与 updateRecords()方法，代码如下所示。

```
1.  public void uploadBook(Book book) {
2.      int count = bookDao.addBook(book);
3.      if (count == 1) {
4.          log.info("uploading successful!");
5.      } else {
6.          log.info("uploading failed!");
7.      }
8.  }
9.
10. public void updateRecords(long uploader, long uploadedBook) {
11.     Upload upload = new Upload();
12.     upload.setUploader(uploader);
13.     upload.setUploadedBook(uploadedBook);
14.     Date uploadedDate = new Date();
15.     upload.setUploadedDate(uploadedDate);
16.     uploadDao.addUploadRecord(upload);
17. }
```

从以上代码中可以看出，调用 bookDao 类的 addBook()方法即可添加书籍；调用 uploadDao 类的 addUploadRecord()方法即可添加上传记录。因为以上代码的业务逻辑较简单，在此不讲解代码。下面在 UserService 类中添加 updateUserContribution()方法，代码如下所示。

```
public void updateUserContribution(int addValue, long userID) {
    userDao.updateUserContribution(addValue, userID);
}
```

从以上代码中可以看出，业务层调用数据库层的方法，增加当前用户的贡献值。

13.8 个人信息模块

个人信息模块是已登录的用户才可以访问的模块，此模块包含查看个人信息、修改信息、修改密码、上传历史和反馈建议 5 个功能。本节将详细讲解这些功能的实现方法。

13.8.1 修改信息页的制作

【实战目标】

单击头像，在弹出的快捷框中单击“个人资料”，跳转到个人信息页。个人信息页不涉及后台数据传输，较简单，在此不做讲解。本节将详细讲解修改信息页的制作方法，修改信息页如图 13.42 所示。

图 13.42 修改信息页

从图 13.42 中可以看出，修改信息页分为 3 个部分，分别为标题栏、左侧的导航列表和右侧的修改信息框。标题栏代码的编写可以参考 13.4.1 小节。下面详细讲解左侧导航列表和右侧修改信息框代码的编写。

【实现步骤】

1．编写修改信息页的代码

在 views 文件夹下创建 personInfo.jsp 文件。下面分步讲解个人信息页代码的编写。

（1）在 personInfo.jsp 文件中编写左侧导航列表的代码，核心代码如下所示。

```
1.  <div class="row">
2.      <div class="col-md-2 col-sm-8 col-xs-12 list-group text-center">
3.          <a data-index="#infoShow" class="personMenu list-group-item" href="#">
个人信息</a>
4.          <a data-index="#infoModify" class="personMenu list-group-item" href="#">
修改信息</a>
5.          <a data-index="#pwdModify" class="personMenu list-group-item" href="#">
修改密码</a>
6.          <a data-index="#uploadHistory" class="list-group-item" href="#"
id="getHistory">上传历史</a>
7.          <a data-index="#feedback" class="personMenu list-group-item" href="#">
反馈建议</a>
8.      </div>
9.
10.     <div class="col-md-10 col-sm-10 col-xs-12">
11.         <div class="panel panel-info">
12.             <div class="panel-heading">
13.                 <span>个人信息</span>
14.             </div>
15.             <div class="panel-body">
16.                 <%@include file="subViews/infoShow.jsp"%>
17.                 <%@include file="subViews/infoModify.jsp"%>
18.                 <%@include file="subViews/pwdModify.jsp"%>
```

```
19.             <%@include file="subViews/uploadHistory.jsp"%>
20.             <%@include file="subViews/feedback.jsp"%>
21.         </div>
22.     </div>
23.   </div>
24. </div>
```

以上代码的第3～7行通过 data-index 属性指定单击导航链接后右侧显示的信息框，同时代码的第16～20行引入个人信息页的5个信息框。

（2）编写修改信息框的代码。

在 subViews 文件夹下创建 infoModify.jsp 文件，编写其中的代码，核心代码如下所示。

```
1.  <form class="form-horizontal" action="infoModify" method="post" onsubmit=
"return checkInfo();">
2.    <div class="info-modify col-md-4 col-sm-12 col-xs-12">
3.      <div class="form-group">
4.        <label for="name"
5.              class="control-label col-md-4 col-sm-6 col-xs-12">
6.          昵称
7.        </label>
8.        <div class="col-md-8 col-sm-6 col-xs-12">
9.          <input id="name"
10.               name="name"
11.               class="form-control"
12.               type="text"
13.               value="${user.userName}">
14.       </div>
15.     </div>
16.     <div class="form-group">
17.       <label for="email"
18.             class="control-label col-md-4 col-sm-6 col-xs-12">
19.         邮箱
20.       </label>
21.       <div class="col-md-8 col-sm-6 col-xs-12">
22.         <input id="email"
23.               name="email"
24.               class="form-control"
25.               type="text"
26.               value="${user.email}">
27.       </div>
28.     </div>
29.     <div class="hide">
30.       <input id="avatarValue"
31.             type="text"
32.             name="avatarImg"
33.             value="${user.avatar.avatar_img}">
34.     </div>
35.   </div>
36.   <div class="col-md-8 col-sm-12 col-xs-12">
37.     <div class="current-img col-md-2 col-sm-12 col-xs-12">
38.       <label class="block">当前头像</label>
39.       <img class="img-thumbnail"
40.   src="${pageContext.request.contextPath}/${user.avatar.avatar_img}"
41.           alt="avatar">
```

```
42.         </div>
43.         <div class="img-select col-md-10 col-sm-12 col-xs-12">
44.             <label class="block">默认头像</label>
45.             <div class="img-group">
46.                 <a href="#">
47.                     <img class="img-thumbnail"
src="${pageContext.request.contextPath}/resources/images/avatars/040601.jpg"
48.                         alt="avatar">
49.                 </a>
50.                 <%--图片集--%>
51.                 ...
52.                 <a href="#">
53.                     <img class="img-thumbnail"
54. src="${pageContext.request.contextPath}/resources/images/avatars/0406010.jpg"
55.                         alt="avatar">
56.                 </a>
57.             </div>
58.         </div>
59.     </div>
60.     <div class="col-md-6 text-center">
61.         <button class="btn btn-primary" type="submit">保存修改</button>
62.     </div>
63. </form>
```

在以上代码中，使用<input>标签来创建昵称和邮箱输入框。以上代码的第46～56行渲染头像列表，当单击头像列表中的头像时，切换为该头像。单击“保存修改”按钮，将修改后的个人信息发送给后端。

2．编写修改信息页的业务逻辑代码

在UserController.java文件中添加infoModify()方法，代码如下所示。

```
1.  @RequestMapping(value = "/infoModify")
2.      public String infoModify(
3.                      String name,
4.                      String email,
5.                      String avatarImg,
6.                      HttpSession session) {
7.          log.info("The user is modifying his information!");
8.          UserHelper userHelper =
9.                  (UserHelper) session.getAttribute("userHelper");
10.         User user = new User();
11.         user.setId(userHelper.getId());
12.         user.setUserName(name);
13.         user.setEmail(email);
14.         int avatarId = userService.getAvatarId(avatarImg);
15.         user.setAvatarNum(avatarId);
16.         userService.updateUserInfo(user);
17.         User user1;
18.         user1 = userService.queryById(userHelper.getId());
19.         UserHelper newUserHelper;
20.         newUserHelper =
21.     userService.getLoginUser(user1.getUserCode(), user1.getUserPassword());
22.         session.setAttribute("userHelper", newUserHelper);
23.         return "redirect:/person";
24.     }
```

下面详细讲解以上代码。

（1）代码的第 8 行和第 9 行从 Session 中获取 UserHelper 对象。然后解析其中的信息，并将其封装为需要更新的 User 对象。

（2）代码第 16 行调用 UserService 类的 updateUserInfo()方法，并将封装好的 User 对象传入，更新用户信息。

（3）代码的第 20～22 行创建新的 UserHelper 对象，并将其放入 Session 中。

下面编写 Service 层的代码。在 UserService.java 文件中添加代码，代码如下所示。

```
1.  public void updateUserInfo(User user) {
2.          userDao.updateUserInfo(user);
3.  }
4.  public int getAvatarId(String avatar_img) {
5.          int avatarId = avatarDao.queryByImgPath(avatar_img);
6.          return avatarId;
7.  }
8.  public User queryById(long id) {
9.          User user;
10.         user = userDao.queryById(id);
11.         return user;
12. }
```

在以上代码中，updateUserInfo()方法根据更新后的 User 对象来更新用户信息，getAvatarId()方法根据图片的地址查询图片的 id，queryById()方法通过用户的 id 查询相应的 User 对象。

13.8.2 修改密码页的制作

【实战目标】

单击个人信息页的“修改密码”导航链接，右侧的信息框切换为修改密码框，如图 13.43 所示。

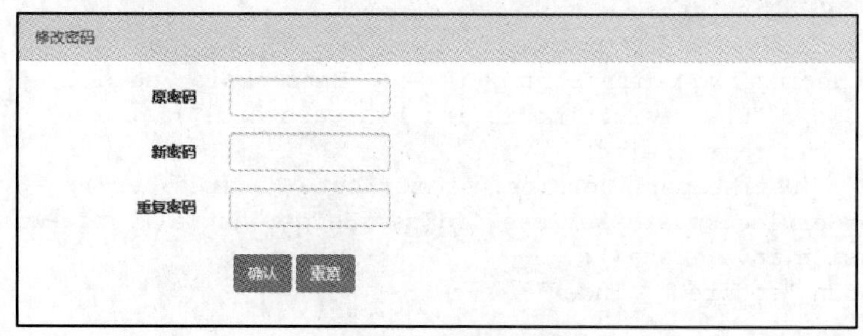

图 13.43　修改密码框

从图 13.43 中可以看出，修改密码框由简单的 3 个输入框、两个按钮等组成。下面介绍修改密码框代码的编写。

【实现步骤】

1. 编写修改密码框的代码

在 subViews 文件夹下创建 pwdModify.jsp 文件，编写其中的代码，核心代码如下所示。

```
1.  <form class="form-horizontal col-md-offset-1"
2.          action="pwdModify" method="post" onsubmit="return checkPwd();">
3.      <div class="form-group">
4.          <label class="control-label col-md-2">原密码</label>
```

```
5.            <div class="col-md-3">
6.                <input id="oldPwd" class="form-control" type="password">
7.            </div>
8.        </div>
9.        <div class="form-group">
10.           <label class="control-label col-md-2">新密码</label>
11.           <div class="col-md-3">
12.               <input id="newPwd" name="newPwd" class="form-control" type=
"password">
13.           </div>
14.       </div>
15.       <div class="form-group">
16.           <label class="control-label col-md-2">重复密码</label>
17.           <div class="col-md-3">
18.               <input id="newPwd2" name="newPwd2" class="form-control" type=
"password">
19.           </div>
20.       </div>
21.       <div class="form-group">
22.           <button class="btn btn-success" type="submit">确认</button>
23.           <button class="btn btn-success" type="reset">重置</button>
24.       </div>
25.   </form>
```

以上代码添加了 3 个输入框，单击“确认”按钮时，输入框中的信息传送给后端。

2. 编写修改密码的业务逻辑代码

向 UserController 类中添加业务代码，由于修改密码的逻辑较简单，在此将 Controller 层和 Service 层的代码一起给出，代码如下所示。

```
1.  //Controller 层的代码
2.  @RequestMapping(value = "/pwdModify")
3.      public String pwdModify(String newPwd, HttpSession session) {
4.          log.info("The user is modifying his password!");
5.          UserHelper userHelper =
6.                  (UserHelper) session.getAttribute("userHelper");
7.          userService.updateUserPassword(userHelper.getId(), newPwd);
8.          session.invalidate();
9.          return "redirect:/index";
10. }
11. //Service 层的代码
12. public void updateUserPassword(long id, String password) {
13.         userDao.updateUserPassword(id, password);
14. }
```

在以上代码中，从 Session 中获取用户信息，从用户信息中获取新的密码，调用 UserService 类的 updateUserPassword()方法将用户的密码更新。

13.8.3 上传历史页的制作

【实战目标】

单击个人信息页的“上传历史”导航链接，右侧的信息框切换为上传历史框，如图 13.44 所示。

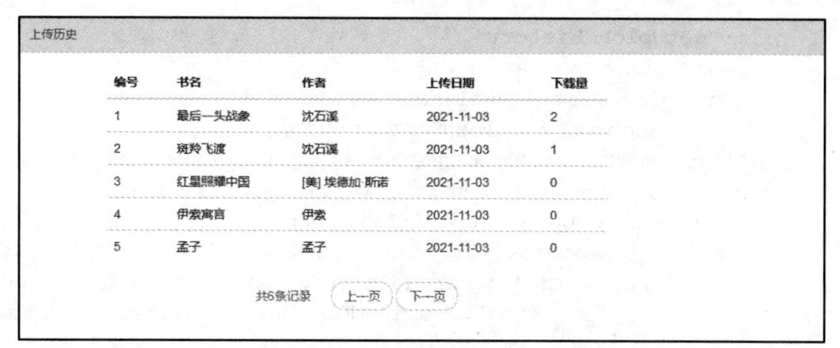

图 13.44　上传历史框

从图 13.44 中可以看出，上传历史框由一个表格和分页按钮等组成。下面讲解上传历史框代码的编写。

【实现步骤】

1．编写上传历史框的代码

在 subViews 文件夹下创建 uploadHistory.jsp 文件，编写其中的代码，核心代码如下所示。

```
1.  <table class="table">
2.          <thead>
3.          <tr>
4.              <th class="col-md-1">编号</th>
5.              <th class="col-md-2">书名</th>
6.              <th class="col-md-2">作者</th>
7.              <th class="col-md-2">上传日期</th>
8.              <th class="col-md-1">下载量</th>
9.          </tr>
10.         </thead>
11.         <tbody>
12.         </tbody>
13.     </table>
14.     <div>
15.         <ul class="pager">
16.             <span class="text-primary" style="padding-right: 20px"></span>
17.             <li id="prePage"><a href="#">上一页</a></li>
18.             <li id="nextPage"><a href="#">下一页</a></li>
19.         </ul>
20.     </div>
```

在以上代码中，使用<table>标签创建表格，字段创建完成后利用 JavaScript 请求后端数据，代码如下所示。

```
1.  $(".row .list-group #getHistory").on({
2.      click: function() {
3.          var $target = $(this).data("index"),
4.              $text = $(this).html();
5.          $(".row .panel-body .index").addClass("hide"),
6.              $($target).removeClass("hide"),
7.              $span.html($text),
8.          0 == getHistory && $.ajax({
9.              method: "GET",
```

```
10.          url: "getUploadHistory",
11.          success: function(data) {
12.              var uploadList = eval(data.uploadList);
13.              uploadRecords = uploadList.length,
14.                  console.log("uploadRecords:" + uploadRecords);
15.              for (var i = 0; i < uploadList.length; i++) {
16.                  var uploadObj = uploadList[i];
17.                  $tbody.append("<tr><td>" + uploadObj.id + "</td> <td>" +
uploadObj.bookTitle + "</td> <td>" + uploadObj.bookAuthor + "</td> <td>" +
uploadObj.uploadedDate + "</td> <td>" + uploadObj.downloadTimes + "</td></tr>")
18.              }
19.              pageTotal = uploadRecords / 5 == 0 ? parseInt(uploadRecords / 5) :
parseInt(uploadRecords / 5) + 1,
20.                  changePage(currentPage),
21.                  $("#uploadHistory .pager span").text("共" + uploadRecords + "
条记录"),
22.                  getHistory = 1
23.          },
24.          error: function() {
25.              $("#uploadHistory .pager span").text("共0条记录"),
26.                $("#uploadHistory .table tbody").append("无法获取上传记录! ")
27.          }
28.      })
29.  }
30. })
```

以上代码为“上传历史”导航链接绑定了响应事件，当单击“上传历史”导航链接时，以上代码的第10行发送请求到后端，当请求得到上传历史数据后，将历史数据拼接到表格内。

2．编写查看上传历史记录的业务逻辑代码

在UserController类中添加业务代码，因为该功能的业务逻辑较简单，在此直接将Controller层的和Service层的代码一起给出，代码如下所示。

```
1.  //Controller层的代码
2.  @RequestMapping(value = "/getUploadHistory")
3.  @ResponseBody
4.  public Map<String, Object> getUploadHistory(HttpSession session) {
5.      Map<String, Object> resultMap = new HashMap<String, Object>();
6.      UserHelper userHelper = (UserHelper) session.getAttribute("userHelper");
7.      List<UploadHelper> uploadHelperList =
8.           userService.getUploadedBook(userHelper.getId());
9.      resultMap.put("uploadList", uploadHelperList);
10.     log.info("you are looking up the uploaded books");
11.     return resultMap;
12. }
13. //Service层的代码
14. public List<UploadHelper> getUploadedBook(long userId) {
15.         List<UploadHelper> uploadHelperList = new ArrayList<UploadHelper>();
16.         List<Upload> uploadList;
17.         uploadList = uploadDao.queryByUserId(userId);
18.         for (int i = 0; i < uploadList.size(); i++) {
19.             UploadHelper upload = new UploadHelper();
20.             upload.setId(i + 1);
```

```
21.             Book book =
22.                 bookDao.queryById(uploadList.get(i).getUploadedBook());
23.             upload.setBookTitle(book.getBook_title());
24.             upload.setBookAuthor(book.getBook_author());
25.             SimpleDateFormat dateFormat = new SimpleDateFormat("yyyy-MM-dd");
26.             String uploadDate =
27.                 dateFormat.format(uploadList.get(i).getUploadedDate());
28.             upload.setUploadedDate(uploadDate);
29.             upload.setDownloadTimes(book.getDownload_times());
30.             uploadHelperList.add(upload);
31.         }
32.         return uploadHelperList;
33. }
```

在以上代码中，Controller 层代码的作用是从 Session 中获取用户信息，从用户信息中获取用户 id，根据此 id 查询用户的上传历史记录并将其返回。Service 层代码的 getUploadedBook()方法首先根据用户 id 查询此用户上传的所有图书，然后循环查询每本图书的具体信息，最后将查询到的图书信息返回。

13.8.4 反馈建议页的制作

【实战目标】

单击个人信息页的“反馈建议”导航链接，右侧的信息框切换为反馈建议框，如图 13.45 所示。

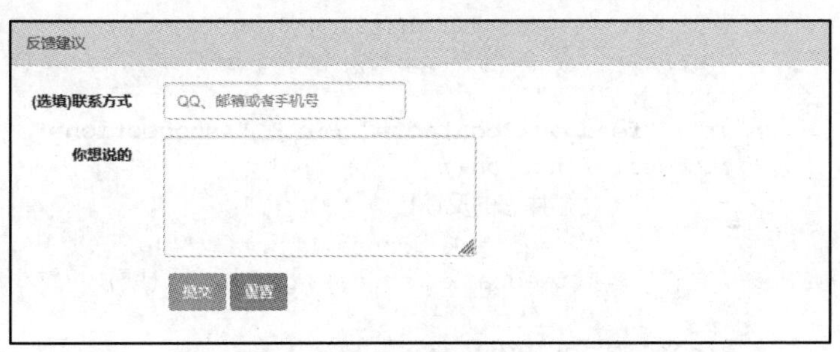

图 13.45 反馈建议框

从图 13.45 中可以看出，反馈建议框由一个输入框和一个文本框等组成。下面讲解反馈建议框代码的编写。

【实现步骤】

1. 编写反馈建议框的代码

在 subViews 文件夹下创建 feedback.jsp 文件，编写其中的代码，核心代码如下所示。

```
1.  <form id="feedbackForm" class="form-horizontal">
2.      <div class="form-group">
3.          <label class="control-label col-md-2">(选填)联系方式</label>
4.          <div class="col-md-4">
5.              <input class="contact form-control"
6.                      type="text"
7.                      placeholder="QQ、邮箱或者手机号">
8.          </div>
9.      </div>
10.
11.     <div class="form-group">
```

```
12.         <label class="control-label col-md-2">你想说的</label>
13.         <div class="col-md-5">
14.             <textarea class="suggestion form-control" rows="4"
15. placeholder="请在此写下你的想法或建议，以及你所遇到的各种问题（不要超过200个字）。">
16.             </textarea>
17.         </div>
18.     </div>
19.
20.     <div class="form-group">
21.         <button id="feedbackSubmit" class="btn btn-warning" type="button">
22.                 提交
23.         </button>
24.         <button class="btn btn-info" type="reset">重置</button>
25.     </div>
26. </form>
```

与上传历史页相同，反馈建议页也由 JavaScript 控制“提交”按钮，代码如下所示。

```
1.  function feedback() {
2.      var e = ($("#feedback #feedbackForm"), $("#feedback #feedbackSubmit"));
3.      e.on({
4.          click: function() {
5.              var e = $("#feedback .contact").val(),
6.                  o = $("#feedback .suggestion").val();
7.              o.length > 400 && (o = o.substring(0, 400)),
8.                  0 == o.length ? $("#feedback .suggestion").focus() : $.ajax({
9.                      method: "GET",
10.                     url: "feedback?contact=" + e + "&suggestion=" + o,
11.                     success: function() {
12.                         alert("感谢您的反馈！"),
13.                             $("#feedback .contact").val(""),
14.                             $("#feedback .suggestion").val("")
15.                     },
16.                     error: function() {
17.                         alert("提交失败，请检查网络连接！")
18.                     }
19.                 })
20.         }
21.     })
22. }
```

以上代码的第2～4行为“提交”按钮绑定了监听事件，当单击“提交”按钮时，立即筛选提交的值，并将数据提交到后端。

2. 编写反馈建议的业务逻辑代码

在 UserController 类中添加业务逻辑代码，因该功能的业务逻辑较简单，在此直接将 Controller 层和 Service 层的代码一起给出，代码如下所示。

```
1.  //Controller 层的代码
2.  @RequestMapping(value = "/feedback")
3.      @ResponseBody
4.      public void feedback(
5.                  String contact,
6.                  String suggestion,
```

```
7.                 HttpSession session) {
8.         UserHelper userHelper =
9.               (UserHelper) session.getAttribute("userHelper");
10.        userService.addFeedback(userHelper.getId(), contact, suggestion);
11.        log.info("you are posting the suggestion!");
12.        log.info("contact:" + contact);
13.        log.info("suggestion:" + suggestion);
14.    }
15. //Service 层的代码
16. public void addFeedback(long userId, String contact, String suggestion) {
17.        Feedback feedback = new Feedback();
18.        feedback.setLoginedUser(userId);
19.        feedback.setContact(contact);
20.        feedback.setSuggestion(suggestion);
21.        feedback.setPostTime(new Timestamp(new Date().getTime()));
22.        feedbackDao.addFeedback(feedback);
23.    }
```

在以上代码中，首先从 Session 中获取当前用户的 id，然后将前端传入的联系方式与建议封装到 FeedBack 类中，最后将 FeedBack 类添加到数据库中。

13.9 后台管理模块

敛书网具有后台管理模块，此模块具有书籍管理、用户管理和反馈处理 3 个功能。本节将详细讲解这些功能的实现方法。

13.9.1 书籍管理页的制作

【实战目标】

访问 bookManage 控制器，页面跳转到首页并打开登录对话框，登录后即可进入后台管理系统。因为登录对话框的业务逻辑较简单，在此不做讲解。进入后台管理系统首先打开的页面是书籍管理页，如图 13.46 所示。

图 13.46　书籍管理页

从图 13.46 中可以看出，整个页面分为两个部分，分别是后台管理系统标题栏和书籍管理列表。本节将带领读者完成书籍管理页的制作。

【实现步骤】

1. 编写书籍管理页的代码

在 backstage 文件夹下创建 bookManage.jsp 文件，分类编写其中的代码。

（1）编写后台管理系统标题栏的代码，核心代码如下所示。

```
1.  <div class="container container-fluid">
2.          <div class="navbar-header">
3.              <a class="navbar-brand" href="#">
4.                  <span class="h3 text-success">敛书网</span>
5.              </a>
6.          </div>
7.          <div class="nav">
8.              <ul class="nav navbar-nav h4">
9.                  <li class="active"><a href="bookManage">书籍管理</a></li>
10.                 <li><a href="userManage">用户管理</a></li>
11.                 <li><a href="feedbackManage">反馈处理</a></li>
12.             </ul>
13.             <div id="user" class="pull-right">
14.                 <h4 class="text-primary">欢迎进入敛书网后台管理系统!  
15.                     <span class="glyphicon glyphicon-user">admin</span></h4>
16.             </div>
17.         </div>
18. </div>
```

在以上代码中，使用<ul>标签和<span>标签创建整个标题栏。

（2）编写书籍管理列表的代码，第一步编写左侧 3 个按钮对应的代码，核心代码如下所示。

```
1.  <div class="searchBy col-md-3 col-sm-3 col-xs-12">
2.      <div id="searchByDate">
3.          <button type="button" class="btn btn-info"
4.                  data-toggle="collapse" data-target="#byDate">
5.              根据上传日期查询
6.          </button>
7.          <div id="byDate" class="collapse">
8.              <form class="searchForm form-horizontal"
9.                    action="searchBookByDays" method="get">
10.                 <div class="input-group">
11.                     <select name="days" class="form-control">
12.                         <option value="1">一天内</option>
13.                         <option value="7">一周内</option>
14.                         <option value="30">一月内</option>
15.                     </select>
16.                     <span class="input-group-btn">
17.                     <button class="btn btn-default" type="submit">
18.                         查询
19.                     </button>
20.                 </span>
21.                 </div>
22.             </form>
```

```
23.           </div>
24.       </div>
25.       <div id="searchByBook">
26.           <button type="button" class="btn btn-success"
27.                   data-toggle="collapse" data-target="#byBook">
28.               根据书籍名称查询
29.           </button>
30.           <div id="byBook" class="collapse">
31.               <form class="searchForm form-horizontal"
32.                     action="searchBookByTitle" method="GET">
33.                   <div class="input-group">
34.                       <input name="title" class="form-control" type="text"
35.                           placeholder="请填写书籍名称">
36.                       <span class="input-group-btn">
37.                           <button class="btn btn-default"
38.                                   type="submit">
39.                               查询
40.                           </button>
41.                       </span>
42.                   </div>
43.               </form>
44.           </div>
45.       </div>
46.       <div id="searchByUser">
47.           <button type="button" class="btn btn-primary"
48.                   data-toggle="collapse" data-target="#byUser">
49.               根据上传用户查询
50.           </button>
51.           <div id="byUser" class="collapse">
52.               <form class="searchForm form-horizontal"
53.                     action="searchBookByUser" method="GET">
54.                   <div class="input-group">
55.                       <input name="userId" class="form-control" type="text"
56.                               placeholder="请填写上传用户编号">
57.                       <span class="input-group-btn">
58.                           <button class="btn btn-default" type="submit">
59.                               查询
60.                           </button>
61.                       </span>
62.                   </div>
63.               </form>
64.           </div>
65.       </div>
66. </div>
```

以上代码使用<form>包裹<input>标签，当单击“查询”按钮时，将需要筛选的数据发送给后端，后端进行模糊查询，将查询到的数据返回给前端，前端接收数据并将其添加进书籍列表中。

第二步编写书籍列表的代码，核心代码如下所示。

```
1.  <div id="bookList" class="col-md-9 col-sm-9 col-xs-12">
2.      <table class="table table-hover">
3.          <thead>
4.          <tr>
5.              <th class="col-md-2 col-sm-2 col-xs-2">编号</th>
```

```
6.            <th class="col-md-2 col-sm-2 col-xs-2">书名</th>
7.            <th class="col-md-2 col-sm-2 col-xs-2">作者</th>
8.            <th class="col-md-2 col-sm-2 col-xs-2">上传者</th>
9.            <th class="col-md-2 col-sm-2 col-xs-2">上传时间</th>
10.           <th class="col-md-2 col-sm-2 col-xs-2">操作</th>
11.        </tr>
12.        </thead>
13.        <tbody>
14.        <c:forEach items="${bookList}" var="book" >
15.           <tr>
16.              <td class="id">${book.id}</td>
17.              <td class="title">${book.title}</td>
18.              <td class="author">${book.author}</td>
19.              <td class="uploader">${book.uploader}</td>
20.              <td class="uploadedDate">
21. <fmt:formatDate value="${book.uploadedDate}" pattern="yyyy-MM-dd"/>
22.              </td>
23.              <td>
24.                 <button class="btn btn-xs btn-danger">删除</button>
25.              </td>
26.           </tr>
27.        </c:forEach>
28.        </tbody>
29.     </table>
30. </div>
```

查询得到的数据通过以上代码在书籍列表中进行展示，代码的第 14～17 行使用<c:forEach>标签以循环的方式创建数据列表，将查询到的数据放到<td>标签中。

2．编写书籍管理的业务逻辑代码

因为删除功能较简单，在此不讲解。下面讲解 3 个按钮筛选功能的实现。

从前面的代码的 3 个表单中可以看出，此处需要编写 3 个控制器。在 controller 文件夹下创建 BackStageController.java 文件，编写其中的代码，核心代码如下所示。

```
1.  @RequestMapping(value = "/searchBookByDays")
2.  public String searchBookByDays(int days,
3.                               Model model, HttpSession session) {
4.     if (null == session.getAttribute("status")) {
5.        return "backstage/adminLogin";
6.     }
7.     List<doBookHelper> bookList;
8.     bookList = backStageService.getBooksByDays(days);
9.     model.addAttribute("bookList", bookList);
10.    return "backstage/bookManage";
11. }
12.
13. @RequestMapping(value = "/searchBookByTitle")
14. public String searchBookByTitle(String title,
15.                               Model model, HttpSession session) {
16.    if (null == session.getAttribute("status")) {
17.       return "backstage/adminLogin";
18.    }
19.    List<doBookHelper> bookList;
```

```
20.     bookList = backStageService.getBooksByTitle(title);
21.     model.addAttribute("bookList", bookList);
22.     return "backstage/bookManage";
23. }
24.
25. @RequestMapping(value = "/searchBookByUser")
26. public String searchBookByUser(long userId, Model model,
27.                                HttpSession session) {
28.     if (null == session.getAttribute("status")) {
29.         return "backstage/adminLogin";
30.     }
31.     List<doBookHelper> bookList;
32.     bookList = backStageService.getBooksByUserId(userId);
33.     model.addAttribute("bookList", bookList);
34.     return "backstage/bookManage";
35. }
```

以上代码中的 3 个控制器的基本逻辑相同。每个控制器均从 Session 中获取登录状态，判断用户是否登录，如果没有登录，跳转到登录页面；如果已经登录，调用相应的 Service 方法进行查询。最后将查询结果封装为 List 集合并返回给前端。

下面编写 Service 层的代码，核心代码如下所示。

```
1.  public List<doBookHelper> getBooksByDays(int days) {
2.      List<doBookHelper> doBookHelperList;
3.      List<Upload> uploadList;
4.      if (days == 30) {
5.          uploadList = uploadDao.searchByThirtyDays();
6.      } else if (days == 7) {
7.          uploadList = uploadDao.searchBySevenDays();
8.      } else {
9.          uploadList = uploadDao.searchByToday();
10.     }
11.     doBookHelperList = iteratorUploadList(uploadList);
12.     return doBookHelperList;
13. }
14.
15. public List<doBookHelper> getBooksByTitle(String title) {
16.     List<doBookHelper> doBookHelperList = new ArrayList<doBookHelper>();
17.     List<Book> bookList = bookDao.searchBookByTitle(title);
18.     for (Book book : bookList) {
19.         doBookHelper bookHelper = new doBookHelper();
20.         bookHelper.setId(book.getId());
21.         bookHelper.setTitle(book.getBook_title());
22.         bookHelper.setAuthor(book.getBook_author());
23.         Upload upload = uploadDao.queryByBookId(book.getId());
24.         bookHelper.setUploader(upload.getUploader());
25.         bookHelper.setUploadedDate(upload.getUploadedDate());
26.         doBookHelperList.add(bookHelper);
27.     }
28.     return doBookHelperList;
29. }
30.
31. public List<doBookHelper> getBooksByUserId(long userId) {
32.     List<doBookHelper> doBookHelperList = new ArrayList<doBookHelper>();
33.     List<Upload> uploadList = uploadDao.queryByUserId(userId);
```

```
34.     Book book;
35.     for (Upload upload : uploadList) {
36.         book = bookDao.queryById(upload.getUploadedBook());
37.         doBookHelper bookHelper = new doBookHelper();
38.         bookHelper.setId(book.getId());
39.         bookHelper.setTitle(book.getBook_title());
40.         bookHelper.setAuthor(book.getBook_author());
41.         bookHelper.setUploader(upload.getUploader());
42.         bookHelper.setUploadedDate(upload.getUploadedDate());
43.         doBookHelperList.add(bookHelper);
44.     }
45.     return doBookHelperList;
46. }
```

在以上代码中，getBooksByTitle()和 getBooksByUserId()方法均通过模糊查询获取相应数据，将其封装为 Helper 对象并返回。因为查询语句较简单，Mapper 层的代码不讲解。getBooksByDays()方法通过判断控制器传入的天数，执行相应的方法，关于通过天数查询的 Mapper 层的核心代码如下所示。

```
1.  <select id="searchByToday" resultMap="uploadMap">
2.      select
3.      <include refid="upload_columns"/>
4.      from upload where TO_DAYS(uploadedDate)=TO_DAYS(NOW())
5.      order by id desc
6.  </select>
7.
8.  <select id="searchBySevenDays" resultMap="uploadMap">
9.      select
10.     <include refid="upload_columns"/>
11.     from upload
12.     where DATE_SUB(CURDATE(), INTERVAL 7 DAY) = date(uploadedDate)
13.     order by id desc
14. </select>
15.
16. <select id="searchByThirtyDays" resultMap="uploadMap">
17.     select
18.     <include refid="upload_columns"/>
19.     from upload
20.     where DATE_SUB(CURDATE(), INTERVAL 30 DAY) = date(uploadedDate)
21.     order by id desc
22. </select>
```

从以上代码中可以看出，searchByToday 查询语句通过 TO_DAYS()方法查询当天上传的所有图书。searchBySevenDays 和 searchByThirtyDays 查询语句分别通过 DATE_SUB()方法查询最近 7 天和最近 30 天内上传的所有图书。

13.9.2 用户管理页的制作

【实战目标】

单击后台管理系统标题栏中的“用户管理”，即可跳转到用户管理页。用户管理页分为两个部分，分别是后台管理系统标题栏和用户管理列表。用户管理列表如图 13.47 所示。

该页面的标题栏同书籍管理页的标题栏相同，本小节将带领读者完成用户管理列表代码的编写。

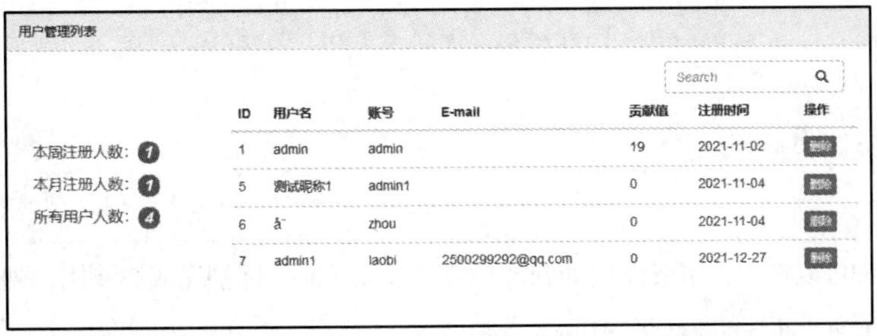

图 13.47 用户管理列表

【实现步骤】

1. 编写用户管理列表的代码

在 backstage 文件夹下创建 userManage.jsp 文件，分类编写其中的代码。

（1）编写左侧的人数统计功能的代码，核心代码如下所示。

```
1.  <div class="col-md-3 col-sm-3 col-xs-12">
2.     <p class="text-info">本周注册人数：<i class="badge">${weekUser}</i> </p>
3.     <p class="text-primary">
4.          本月注册人数：<i class="badge">${monthUser}</i>
5.     </p>
6.     <p class="text-success">
7.          所有用户人数：<i class="badge">${totalUser}</i>
8.     </p>
9.  </div>
```

从以上代码中可以看出，使用<p>标签和简单的 JSP 取值即可实现此功能。

（2）编写右侧用户列表的代码，核心代码如下所示。

```
1.  <table class="table table-hover">
2.     <thead>
3.     <tr>
4.        <th>ID</th>
5.        <th>用户名</th>
6.        <th>账号</th>
7.        <th>E-mail</th>
8.        <th>贡献值</th>
9.        <th>注册时间</th>
10.       <th>操作</th>
11.    </tr>
12.    </thead>
13.    <tbody>
14.       <c:forEach items="${userList}" var="user">
15.          <tr>
16.             <td class="id">${user.id}</td>
17.             <td class="userName">${user.userName}</td>
18.             <td>${user.userCode}</td>
19.             <td>${user.email}</td>
20.             <td>${user.contribution}</td>
21.             <td><fmt:formatDate value="${user.creationDate}"
22.                          pattern="yyyy-MM-dd"/></td>
23.             <td>
```

```
24.                 <button class="btn btn-xs btn-danger">删除</button>
25.              </td>
26.           </tr>
27.        </c:forEach>
28.     </tbody>
29. </table>
```

从以上代码中可以看出，主要使用<table>标签和<c:forEach>标签完成右侧用户列表的相应功能。

2．编写用户管理列表的业务逻辑代码

在 BackStageController.java 文件中添加控制器代码，当后台管理系统标题栏中的“用户管理”被单击时，执行此控制器代码。核心代码如下所示。

```
1.  @RequestMapping(value = "/userManage")
2.  public String userManage(Model model, HttpSession session) {
3.      if (null == session.getAttribute("status")) {
4.          return "backstage/adminLogin";
5.      }
6.      Map<String, Object> resultMap;
7.      resultMap = backStageService.getUserByContribution();
8.      List<User> userList = (List<User>) resultMap.get("userList");
9.      int totalUser = (Integer) resultMap.get("totalUser");
10.     int weekUser = (Integer) resultMap.get("weekUser");
11.     int monthUser = (Integer) resultMap.get("monthUser");
12.     model.addAttribute("userList", userList);
13.     model.addAttribute("weekUser", weekUser);
14.     model.addAttribute("monthUser", monthUser);
15.     model.addAttribute("totalUser", totalUser);
16.     return "backstage/userManage";
17. }
```

以上代码首先判断用户是否登录，然后访问 Servlet 层的代码，获取用户表单数据和注册人数，最后将数据返回给前端。Service 层的核心代码如下所示。

```
1.  public Map<String, Object> getUserByContribution() {
2.         Map<String, Object> resultMap = new HashMap<String, Object>();
3.         List<User> userList = userDao.queryUserByContribution();
4.         resultMap.put("userList", userList);
5.         resultMap.put("totalUser", userDao.queryUserNumber());
6.         resultMap.put("weekUser", userDao.queryUserNumberByWeek());
7.         resultMap.put("monthUser", userDao.queryUserNumberByMonth());
8.         return resultMap;
9.  }
```

在以上代码中，调用 UserDao 类的 queryUserByContribution()方法获取按照贡献值排序的用户列表，调用 UserDao 类的 queryUserNumber()方法获取用户总数，调用 UserDao 类的 queryUserNumberByWeek()方法获取本周新注册的用户总数，调用 UserDao 类的 queryUserNumberByMonth()方法获取本月新注册的用户总数，然后将这些数据封装为 Map 集合并返回给控制器。

13.9.3 反馈处理页的制作

【实战目标】

单击后台管理系统的标题栏中的“反馈处理”，即可跳转到反馈处理页。其中的反馈信息管理列表如图 13.48 所示。

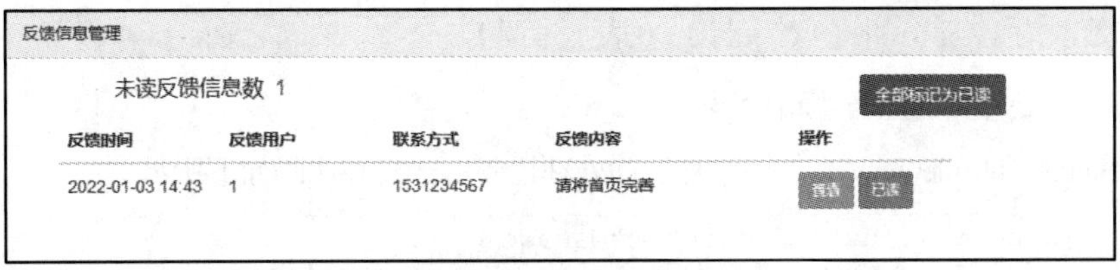

反馈信息管理

未读反馈信息数 1　　全部标记为已读

反馈时间	反馈用户	联系方式	反馈内容	操作
2022-01-03 14:43	1	1531234567	请将首页完善	查看 已读

图 13.48　反馈信息管理列表

从图 13.48 中可以看出，反馈信息管理列表分为未读反馈信息数、“全部标记为已读”按钮和反馈列表。未读反馈信息数和“全部标记为已读”按钮的前端设计此处不讲解。本节将带领读者完成反馈列表的制作。

【实现步骤】

1．编写反馈列表的代码

在 backstage 文件夹下创建 feedbackManage.jsp 文件，向其中添加代码，核心代码如下所示。

```
1.  <table class="table table-hover">
2.      <thead>
3.      <tr>
4.          <th class="col-md-2 col-sm-2">反馈时间</th>
5.          <th class="col-md-2 col-sm-2">反馈用户</th>
6.          <th class="col-md-2 col-sm-2">联系方式</th>
7.          <th class="col-md-3 col-sm-3">反馈内容</th>
8.          <th class="col-md-2 col-sm-2">操作</th>
9.      </tr>
10.     </thead>
11.     <tbody>
12.     <c:forEach items="${feedbackList}" var="feedback">
13.         <tr>
14.             <td class="id" style="display: none">${feedback.id}</td>
15.             <td>
16. <fmt:formatDate value="${feedback.postTime}" pattern="yyyy-MM-dd HH:mm"/>
17.             </td>
18.             <td>${feedback.loginedUser}</td>
19.             <td>${feedback.contact}</td>
20.             <td class="suggestion">${feedback.suggestion}</td>
21.             <td>
22.                 <button class="btn btn-sm btn-info" type="button">
23.                     查看
24.                 </button>
25.                 <button class="btn btn-sm btn-success" type="button">
26.                     已读
27.                 </button>
28.             </td>
29.         </tr>
30.     </c:forEach>
31.     </tbody>
32. </table>
```

以上代码用于创建反馈列表，在列表中有“查看”和“已读”按钮。当单击“查看”按钮时，仅将反馈内容以弹窗形式展示，并无后端交互操作。当单击“已读”按钮时，将数据库中 feedback 表的

status 字段的值改为 1，表示此条反馈已读（0 表示未读）。

2．编写反馈处理页的业务逻辑代码

在 BackStageController.java 文件中添加反馈处理控制器，核心代码如下所示。

```
1.  @RequestMapping(value = "/feedbackManage")
2.  public String feedbackManage(Model model, HttpSession session) {
3.      if (null == session.getAttribute("status")) {
4.          return "backstage/adminLogin";
5.      }
6.      List<Feedback> feedbackList = backStageService.getFeedback();
7.      model.addAttribute("feedbackNum", feedbackList.size());
8.      model.addAttribute("feedbackList", feedbackList);
9.      return "backstage/feedbackManage";
10. }
11. @RequestMapping(value = "/setRead")
12. @ResponseBody
13. public void setFeedbackRead(int feedbackId) {
14.     backStageService.setOneFeedbackRead(feedbackId);
15.  }
16.
17. @RequestMapping(value = "/setAllRead")
18. public String setAllFeedbackRead() {
19.         backStageService.setAllFeedbackRead();
20.         return "redirect:feedbackManage";
21. }
```

上述代码包含 3 个方法，feedbackManage()方法负责查询出反馈总数与未读反馈数量，setFeedbackRead()方法负责将对应 id 的反馈标记为已读，feedbackManage()方法负责将所有反馈标记为已读。

13.10 本章小结

只有把理论知识同具体实际相结合，才能正确回答实践提出的问题，扎实提升读者的理论水平与实战能力。本章先从项目设计和功能结构等方面对项目案例（敛书网）进行了介绍，然后通过数据表的设计讲解了整个项目的实体类，最后以模块的方式讲解了敛书网的具体业务功能。读者在学习本章后需要掌握 SSM 的框架整合技术，以及框架在项目开发中的具体应用。